ECOFEMINISM
AND GLOBALIZATION

ECOFEMINISM AND GLOBALIZATION

Exploring Culture, Context, and Religion

Edited by
Heather Eaton
and
Lois Ann Lorentzen

ROWMAN & LITTLEFIELD PUBLISHERS, INC.
Lanham • Boulder • New York • Toronto • Oxford

ROWMAN & LITTLEFIELD PUBLISHERS, INC.

Published in the United States of America
by Rowman & Littlefield Publishers, Inc.
A wholly owned subsidiary of The Rowman & Littlefield Publishing Group, Inc.
4501 Forbes Boulevard, Suite 200, Lanham, Maryland 20706
www.rowmanlittlefield.com

PO Box 317
Oxford
OX2 9RU, UK

British Library Cataloguing in Publication Information Available

Library of Congress Cataloging-in-Publication Data

Eaton, Heather, 1956–
Ecofeminism and globalization : exploring culture, context, and religion / edited by
　Heather Eaton and Lois Ann Lorentzen.
　　p. cm.
　Includes bibliographical references and index.
　ISBN 0-7425-2697-6 (alk. paper)—ISBN 0-7425-2698-4 (pbk. : alk. paper)
　1. Ecofeminism. 2. Globalization. I. Lorentzen, Lois Ann, 1952– II. Title.

HQ1194.E38 2003
305.42'01—dc21   2003000833

Printed in the United States of America

⊗™ The paper used in this publication meets the minimum requirements of American
National Standard for Information Sciences—Permanence of Paper for Printed Library
Materials, ANSI/NISO Z39.48-1992.

Contents

Foreword

Ecofeminism and the Challenges of Globalization

Rosemary Radford Ruether

T HIS VOLUME IS AN IMPORTANT EFFORT to evaluate the discourse of ecofeminism, which has emerged worldwide in the last three decades, and to explore its adequacy to the challenges of "globalization"; that is to say, to the vast human misery and degradation of the environment that is being wrought by the Western corporate domination of the world economy. The book brings together both critics and advocates of ecofeminist discourse. There are those who write from a more theoretical perspective and those who delve into concrete cases of the interconnection of women and ecology. There are those who write from a first-world context and others from the "Two-Thirds World," from Africa, India, Mexico, Chile, and Brazil. Some writers are rooted in Christianity; others examine the issue from other religious or philosophical traditions, such as Buddhism. All are interested in interconnections, not only between issues of sexism and ecology, but also between theory and practice, between spirituality and ethical action. As someone who has written about ecofeminism before the word itself was coined—if you count my 1975 book, *New Women, New Earth*, as an ecofeminist classic (as some have named it)—I am honored to have been asked to write the preface to this rich collection of essays.

This interconnection between the subjugation of women and that of subjugated races and classes means that ecofeminism cannot treat women as a univocal category. Women are a gender group within every class and race. That means they share in the privileges or disprivileges of their class and race, while also being inferiorized as women in relation to men within their class and race. But this disprivileging as women vis-à-vis the men of their class and race obviously takes very different forms across classes and races. Women

[handwritten margin notes: PARTICU-LARLY NEGATIVE DEFINITION; GENDER GROUP W/IN CLASS/ RACE; DOUBLE VICTIMIZATION]

servants or slaves experienced much more oppressive lives in every way. Since women of the elite were often their more immediate oppressors, it is hard for such Black or working-class women to see themselves as sharing a common oppression with white elite women. It takes a bit of perspective to recognize that all these women, as well as male servants and workers, are part of one system designed to place different groups in different roles across class and race for the benefit of one master group, elite white males. People today, unaccustomed to such ecofeminist analysis, may be inclined to dismiss it as either exaggerated or passé, as they look at the way in which women of elite classes and races have won their way into something like the privileges of their brothers. Far from making ecofeminism irrelevant, these challenges, to show the complexity of gender within class and race, reveal why ecofeminism must interconnect with the movements against environmental racism and for ecojustice and situate itself in a global context. By environmental racism, I mean those movements among African Americans, Latinos, and indigenous peoples (mostly spearheaded by women of these groups) that are struggling against toxic dumping and environmental pollution that is concentrated particularly in the areas where poor people of color live. Global ecofeminism shows how these patterns of impoverishment of nature and of emiserated humans are interconnected in a worldwide economic system skewed to the benefit of the rich beneficiaries of the market economy.

The reality that women are still the poorest of the poor becomes starkly evident whenever gender is analyzed across class and race worldwide. An essay on women in relation to world population in the *2002 State of the World Report* makes this clear. Two-thirds of the world's 876 million illiterate people are female. In twenty-two African nations and nine Asian nations school enrollment for girls is less than 80 percent that of boys. Only 52 percent of girls stay in school past the fourth grade in these countries. Only about four women per one thousand attend high school, much less college. Even in the United States, where 18 percent of households are headed by women, these households account for one-third of the children living in poverty. Women throughout the world earn significantly less than men, on average between two-thirds and three-fourths, and women account for only 5 percent of the senior staff of the five hundred largest corporations. Only in nine nations are 30 percent or more members of parliament women, while in the Americas women hold only 15 percent of parliamentary seats and only 4 percent in Arab states. Seven African and Arab states have no women members of their legislatures. In many states, women are still legally under the guardianship of their husbands or fathers and have no right to manage property.

Physical abuse shadows women from before birth. Sex-selected abortions, female infanticide, malnutrition, and abuse of female children are common in many nations. In India dowry murders, the killing of wives in order to seek dowries from second wives, continue to happen despite twenty-five years of efforts to expose this practice. Incest, female genital cutting, denial of medical care, early marriage, forced prostitution, and forced labor hang over women's heads worldwide. Girls and women are more likely to be sold into slavery than males, and some 130 million women worldwide have experienced the cutting of their genitals, a practice that continues at the rate of about two million a year.

[handwritten margin note: DOES THIS PRESENT A ONE-SIDED VIEW OF WESTERN OF THESE PRACTICES ??]

Those concerned with both population and environment have recognized that the single factor most likely to both check population expansion and also improve the health and welfare of children in families and care for the local environment is the promotion of the equality of women with men. Americans are likely to assume that women and children share in affluence or poverty pretty much on the same level with the men of their family, but in fact studies continually show that men tend to use the majority of their own assets for themselves, not for the women and children of their own families. Women, by contrast, devote the great majority of the fruits of their own labor, whether in cash or in subsistence labor, to feeding, clothing, and educating their children. Women also do much of the subsistence labor that protects and renews local environments. Thus, increasing women's share in education, income, and power is a major factor in improving the health, welfare, and education of children.

We must be clear that promoting women's equality is not a matter of isolating women from men or children. We need to continually insist that feminism is about converting the relations of patriarchal domination for both women and men into new relationships of mutuality. Not only is this not antifamily, but in fact families, and particularly children, are the first beneficiaries of such restoration of men to caring relations with women and children. Likewise ecofeminism must not be seen as making women the primary caretakers for the local environment, but as bringing men into the work of care for the household and earth that is now borne disproportionately by women. Why is it that it is predominantly women, rather than men and women together, who have led the struggles against toxic environments and who do the bulk of the recycling of wastes?

I agree that ecofeminism among middle-class Western women may be turned into a kind of consumer "spirituality," disconnected from any socioeconomic critique, and thus it may become irrelevant to the struggle against global oppression of the poor and the devastation of the planet. I myself experienced this dichotomy about ten years ago when I was invited by the organizers of a large conference of "new age" psychiatrists to speak on my book *Gaia and God*. The conference attracted thousands of women and men interested in self-cultivation. It was held in an expensive crystal palace hotel outside Washington D.C. My session on Gaia and ecological healing attracted about three hundred people into a crowded room. But as soon as I spoke my first sentence, in which I said that we have to look at the issue of ecology from the perspective of the poorest women of the world, half of this group got up and left. The experience made it graphically clear to me that there is the danger for ecofeminist thought to be turned into a leisure class "spirituality," unconnected to poverty, specifically the poverty of the poorest women of the world. Although this danger of an ecofeminist "spirituality," split from ecofeminism as socioeconomic analysis and struggle against the structures of impoverishment of both women and the earth, is a danger that needs continual critique, there is ample evidence of the power of ecofeminist thought and practice where the two are clearly integrated. I give two examples here of where this is happening.

One is these is the foundational work by Maria Mies and Vandana Shiva, expressed in all their work, particularly their joint volume entitled *Ecofeminism* (Zed, 1993). Maria Mies is a German social critic who has long been involved in movements in Europe and in India that interconnect the critique of patriarchy, poverty, and environment on a world level. Among her books are *Patriarchy and Accumulation on a World Scale* (1986), and *Women: The Last Colony* (1988), with Claudia von Werlhof and Veronica Bennholdt-Thomsen. Vandana Shiva is an Indian feminist critic of science who directs the Research Foundation for Science, Technology, and Natural Resource Policy in Dehradun, India. Her first major book on ecofeminism was the widely read *Staying Alive: Women, Ecology and Development* (Zed, 1989).

In their joint work, *Ecofeminism*, Mies and Shiva engage in a critique of patriarchal ideologies enshrined in both Western science and the "development" establishment. They reveal in concrete detail how the policies promoted by the ideologies of these institutions are impoverishing the planet and the majority of its peoples, with women and children as primary victims. Here ecofeminist critique and activism stand in direct relation to the issues of globalization. Neither Mies nor Shiva are theologians or much interested in spirituality, although Shiva does speak of the Hindu vision of Shakti, or female cosmic life energy, and this worldview is seen as promoting a different understanding of our relation to the planet as earth family.

Humans must recognize themselves anew as embedded in the Earth family. By contrast, the recent book by Starhawk, a feminist Wiccan priestess, *Webs of Power: Notes from the Global Uprising* (New Society, 2002), shows how a newly emergent Earth-based spirituality can be the direct underpinnings of the global struggle against the World Bank and the World Trade Organization (WTO) as well as challenging the global power holders in venues of their own global gatherings.

Not only are ecofeminism and a critique of globalization deeply interconnected, indeed are one, in these volumes, but they also show that spirituality and concrete action must be one in the struggles on behalf of the Earth and its peoples against global domination. These connections of ecofeminism as spirituality and as ethical practice in the actual struggles against corporate violence need to be continually rediscovered and made evident. This volume can contribute another piece to that work.

Acknowledgments

THIS BOOK GREW OUT OF CONVERSATIONS between Heather and Lois following a panel on ecofeminism at the American Academy of Religion's Annual Meeting in 1998 in Orlando, Florida. We had been grappling with ecofeminism for some time but wanted a larger and more concrete conversation about its potential and its weaknesses. We are delighted that others find this exploration of and challenge to ecofeminism relevant to their work. We first want to thank the contributors who responded to our call and joined our efforts to rethink ecofeminism. The authors write from all parts of the world and from many disciplines. Each person struggles to bridge the divides between thought and action, academia and activism. We are grateful to have companions, all trying to make the world more livable, enjoyable, beautiful, and peaceful. Editing *Ecofeminism and Globalization: Culture, Context, and Religion* has been a great privilege.

Lois and Heather thank the Faculty Development Fund of the University of San Francisco for extending funding at a critical juncture in the book's genesis; especially Jennifer Turpin of the Dean's Office of the College of Arts and Sciences. Teresa Walsh, editor extraordinaire, along with editors Rebekah Bloyd, Steven Saum, and Nikki Thompson, provided invaluable help in reviving prose and crafting this volume. Research assistant Calgary Brown gave painstaking help in editing and tracking down copyright permissions, always with grace and humor. We thank Saint Paul's University in Ottawa for funding assistance. Special thanks go to Brian Romer of Rowman & Littlefield who provided prompt and friendly assistance. Lois offers a particular appreciation to Gerardo Marin, as always. Finally, we thank each other for being patient, committed, engaged, and intelligent partners in this project.

Introduction

Heather Eaton and Lois Ann Lorentzen

DEFORESTATION, WARS, MILITARIZATION, and the socioeconomic impoverishment of many of the globe's women challenge all who hope for a more humane world. How best to respond to ecological crises, gender concerns, and increased violence? Many activists and academics are drawn to ecofeminism, an intellectual and activist movement that makes critical connections between the domination of nature and the exploitation of women. Although we have studied ecofeminism for years and have developed ecofeminist analyses, we also have become aware of limitations when ecofeminism confronts particular and concrete problems. How effective are ecofeminist approaches in the face of the critical problems listed above? Are connections between women and nature the same in every culture, religion, and context? How does ecofeminism confront the many faces of globalization? These questions, offered from the perspective of those most affected by socioeconomic globalization, contextualize the debates about ecofeminism presented in this book.

What Is Ecofeminism?

Ecofeminism encompasses a variety of theoretical, practical, and critical efforts to understand and resist the interrelated dominations of women and nature. The term *ecofeminism* originated with French feminist Françoise d'Eaubonne in 1974 and became a useful tool to assess interconnections between women's oppression and the ecological crisis. Ecofeminism in North America has roots in antinuclear, antiwar, environmental, and lesbian-feminist activist movements. Numerous

— 1 —

conferences and popular ecofeminist publications in the 1980s were precursors
to a burgeoning of academic ecofeminist writings, especially in the fields of philosophy, theology, and religious studies. Analyses of critical links between militarism, sexism, classism, racism, and environmental destruction became central
to ecofeminist thought and action. Ecofeminism now reflects the concerted efforts of women trying to integrate their personal, ecological, and sociopolitical
concerns.

Although ecofeminism explores a range of women/nature interconnections, three claims seem central—the empirical, the conceptual (cultural/
symbolic), and the epistemological. The empirical claim is that environmental problems disproportionately affect women in most parts of the world. The
United Nations as early as 1989 observed, "It is now a universally established
fact that it is the woman who is the worst victim of environmental destruction. The poorer she is, the greater is her burden."[1] The increased burdens
women face result not only from environmental deterioration; the sexual division of labor found in most societies considers family sustenance to be
women's work, and women, as primary caregivers, generally bear primary responsibility for the food and the health of family members. Providing fuel,
food, and water for families becomes increasingly difficult with environmental degradation. To make matters worse, economic resources—ownership of
land or commercial businesses—remain inaccessible to most women.
Ecofeminists' empirical claim examines sociopolitical and economic structures that restrict many women's lives to poverty, ecological deprivation, and
economic powerlessness.

A second claim is that women and nature are connected conceptually and
symbolically in Euro-western worldviews. These connections are articulated in
several ways. According to ecofeminists, Euro-western cultures developed ideas
about a world divided hierarchically and dualistically. Dualistic conceptual structures identify women with femininity, the body, sexuality, earth or nature and
materiality; and men with masculinity, the mind, heaven, the supernatural, and
disembodied spirit. Dualisms such as reason/emotion, mind/body, culture/
nature, heaven/earth, and man/woman give priority to the first over the second.
Ecofeminists refer to these pairings as hierarchical dualisms and claim they point
to a logic of domination that is entrenched in Euro-western history and worldviews. Religion, philosophy, science, and cultural symbols reinforce this worldview, making male power over both women and nature appear "natural" and
thus justified. Social patterns, including sexual norms, education, governance,
and economic control, reflect this logic of domination.

The third claim is epistemological, following from connections noted between
women and nature. Since environmental problems affect women most directly
isn't it possible that women possess greater knowledge and expertise that could

Handwritten margin notes: "3 CENTRAL CLAIMS"; "① WOMEN SUFFER MOST FROM ENV. DEGRADATION"; "② WOMEN AND NATURE ARE CONNECTED CONCEPTUALLY IN EURO-WESTERN WORLDVIEWS"

ⓐ

prove useful in finding solutions to pressing environmental problems? Some EPISTE-
claim women possess more knowledge about Earth systems than men and thus NOLOGISM
should be "epistemologically privileged." For example, in many parts of the
world, women are the land's custodians and have greater agricultural knowledge
than men. Thus, according to ecofeminists, these women are in a good position
to aid in creating new practical and intellectual ecological paradigms. Women,
then, are best equipped to address local environmental problems. For a few
ecofeminists this epistemological privilege is more than a pragmatic claim however. Some see women as "closer" to nature/earth than men and possessing in- ✱ ✻
nate traits of caring, community building, nonviolence, and Earth sensitivity.
This type of women-nature essentialism may be found in some cultures and contexts; however, most ecofeminists consider the connections between women and
nature to be based in theory and in conceptual frameworks rather than in
essence. Where women have expertise in agriculture and ecological systems, it is
not due to their essential nature but to their life experiences.

Ecofeminism, although generally sharing the preceding three claims, is not
one thing. It is, as Karen Warren notes, an umbrella term for a wide variety of
approaches. Ecofeminist theorists differ on foundational assumptions, on the
nature of the relationship between women and the natural world, on ecological paradigms, on feminist approaches, on the roots of environmental crises,
and on goals and the means of achieving them. Ecofeminists may be liberal,
Marxist, socialist, cultural, radical, postmodernist, or ecowomanist. They may
advocate environmental resource management, deep ecology, social ecology,
or new cosmologies in their ecological frameworks. Buddhist, Native American, goddess-worshiping, Hindu, Muslim, Christian, Jewish, and thoroughly
secular versions of ecofeminism exist. Ecofeminist thought and activism occur
in India, Asia, Africa, and Latin America. Regional, ethnic, and cultural
ecofeminisms exist.

If myriad voices, theoretical positions, praxis, and political leanings characterize ecofeminism, then what makes ecological feminism *feminist* is the commit- ✱
ment to the recognition and elimination of male-gender bias and the development of practices, policies, and theories without this bias. What makes it
ecological is a commitment to the valuing and preserving of ecosystems, broadly
understood.[2] Ecofeminism is a textured field of theoretical and experiential insights encompassing different forms of knowledge, embodied in the concrete.

What Is Globalization?

How can we understand globalization? Many claim that globalization is what
most defines this era in human organization. Yet what does it really mean?

Although used frequently, the word has multiple and ambiguous meanings. Globalization can mean: (1) An *economic agenda* that traverses the world, promoting market economies and enhancing trade in the service of capital growth; (2) An *ideology* representing values, cultural norms, and practices, seen by some as a superior worldview and by others as cultural hegemony; (3) A *corporate structure and mechanism* that may supercede the rule of nation-states and challenge or even threaten democracy; (4) A *global village,* the consequence of vast cultural exchanges, communication technologies, transportation, migrations, and a wide array of global interconnections, including the globalization of ideas such as ecofeminism; or (5) A *grassroots globalization* or *globalization from below* as witnessed in anti-globalization or pro-democracy movements emerging in resistance to economic and cultural globalization.

[margin note: DEFINITION OF GLOBALIZATION]

Globalization has many faces, and the authors in this book use the word in a variety of ways. Yet despite this diverse usage, globalization generally refers to the economic and technological agenda that alters basic modes of cultural organization and international exchange in many parts of the world. Heated debates erupt over whether globalization is primarily beneficial or destructive. What is economic globalization achieving? Are the consequences as beneficial as the rhetoric claims? Global monetary institutions and financial elites declare endless benefits from the exchange of capital on a global scale. They praise the increases in export commodities and trade, the participation in global wealth, and what they see as the stabilizing of democratic civic processes. The anti-globalization/pro-democracy voices say that global financial institutions are reordering the flow of capital and wealth within a global and hegemonic economic regime that serves only the interests of the elite and rich. They speak of devastation resulting from globalization including ecological ruin, poorer working conditions, urban deterioration, and increased social violence. Surely both cannot be equally accurate.

Numerous social and political developments accompany economic globalization as governments restructure their services in order to gain access to the global economy. Needed public services such as electricity, water, health, and transportation often are placed in the hands of private companies. The poor find it harder to access these privatized "public" services. With globalization comes what some describe as massive consequences in the area of culture.[3] Values, attitudes, and cultural worldviews are wrapped within the package of globalization. Encounters between the allegedly value-neutral global market and specific cultures with particular values and practices have produced all kinds of clashes, including overt resistance. Societies with public religious structures, values, and social patterns may be deeply challenged as they either absorb or resist the values embedded within globalization processes. Globalization is anything but neutral.

The Problem

Ecofeminism is considered to be a third wave of feminism. For over two decades ecofeminism has expanded its conceptual framework and praxis-based movement. It continues to mature theoretically and grow as an activist response to both ecological crises and gender concerns, while promoting antimilitarism and peace. Like ecology and feminism it is heterogeneous. And although it is considerably developed in both popular movements and academic discourse, ecofeminism remains largely a theoretical conversation. Ecofeminism has, at different times, been labeled as apolitical, essentialist, personalist, and socially unliberatory. The ecofeminism coming from North America has been criticized for ignoring social, cultural, and historical contexts; differences of class, culture, and race among women; and for ignoring the plight of impoverished women from less affluent nations. Some question whether the woman/nature link holds crossculturally. In general these critiques, often from ecofeminists themselves, have been the impetus for the expansion and refinement of ecofeminist theory.

Globalization cannot be ignored as it is a powerful force worldwide. Yet few ecofeminists address globalization.[4] There are at least three reasons for this. First, the largely theoretical discourses linking women and nature, as developed thus far, do not sufficiently address material exclusions resulting from economic forces. Conversation between ecofeminists and economic theorists is minimal. The insistence upon the primacy of a women-nature connection, while illuminating symbolic and cultural constructs, doesn't help us adequately analyze globalization as an extension of patriarchal capitalism. Second, while there are many grassroots activist women's organizations resisting the negative effects of globalization, these activities do not provide the primary data for ecofeminist discourse. Third, an adequate discussion of globalization must include not only an analysis of the economic agenda, its hegemonic impact, and implicit value system, but also the consequences of the erosion of nation-states and the rise of international civic movements. This book asks ecofeminism, and ecofeminists, to seriously grapple with globalization.

Ecofeminism and Globalization: Exploring Religion, Culture, and Context

Ecofeminism and Globalization explores the challenge posed to ecofeminism in an age of globalization. The book presents the work of scholars on the topics of ecofeminism and globalization and is oriented toward examining whether ecofeminism "works" in concrete situations. It offers a combination of approaches: empirical studies, empirically informed essays, theoretical works, and descriptive articles on particular ecofeminist movements. *Ecofeminism and Globalization* utilizes context-specific analyses to understand the particularity of culture, context, and religion, and what occurs in the engagement with the ideas and ideals of ecofeminism. This collection is about a precise question; that is, "does ecofeminism make sense, and can it be effective in concrete, complex situations?" The answers are as varied as the authors.

Ecofeminism as a meaningful crosscultural analysis is spreading, but it is not uniform and does not take shape in the same way in each context. We believe generalizations about globalization or ecofeminism break down when culture, context, religion, and social norms are taken seriously. *Ecofeminism and Globalization* brings reflections from many regions, cultures, and religions of the world, including essays from Taiwan, Mexico, Kenya, Chile, India, Brazil, Canada, England, and the United States. This book shows ecofeminism's effectiveness in some contexts and not in others, with some religions and not with others.

The two opening essays provide horizons in which the questions and challenges of the book are addressed. Mary Mellor gives an overview of the gender/nature relationship and introduces major themes found in ecofeminism and globalization. Heather Eaton analyzes relationships among economic globalization, corporate structures, ecofeminism, and theology.

The second section explores challenges to core premises of ecofeminism on the basis of concrete case studies. Celia Nyamweru offers firsthand accounts of women's use of and attitudes to sacred groves in coastal Kenya. Lois Ann Lorentzen questions basic ecofeminist principles given the practices and cosmologies of indigenous women and men in Chiapas, Mexico. Aruna Gnanadason considers ecofeminism in light of the environmental impact of economic liberalization policies in India.

A third section addresses regional expressions of ecofeminism and specific responses to political forces and globalization. Noël Sturgeon discusses the relationship between ecofeminism and transnational environmental politics from an American viewpoint. Wan-Li Ho describes Taiwanese Buddhist women's environmental movements and their responses to globalization. Mary Judith Ress chronicles the Con-spirando ecofeminist collective and movement in Chile and its response to economic neoliberalism. Ivone Gebara describes the religious and historical connection between the domination of women, animals, and land during and after colonization in Brazil. Masatsugu Maruyama offers a critical interpretation of the similarities and differences between North American ecofeminism and the Japanese indigenous religion Shinto. Greta Gaard gives an account of the history of ecofeminists in the Green political party in the United States.

We hope this volume contributes to the evolution of ecofeminist perspectives. What happens when ecofeminist thought takes context and globalization seriously? It is hoped that theoretical and material concerns come closer together and ecofeminists become politically influential forces in shaping our future(s).

Notes

1. Pamela Philipose, "Women Act: Women and Environmental Protection in India," in *Healing the Wounds: The Promise of Ecofeminism*, ed. Judith Plant (Toronto: Between the Lines, 1989), 67.

2. Karen Warren, ed., *Ecological Feminism* (New York: Routledge, 1994), 1.

3. Peter Berger, "Globalization and Religion," *Globalization and Religion: The Hedgehog Review* 4, 2 (summer 2002): 7–20.

4. The only article to date that we know of is that of Jasmin Sydee and Charson Beder, "Ecofeminism and Globalization: A Critical Appraisal," *Democracy and Nature* (July 2001): 281–302.

I

ECONOMIC GLOBALIZATION, THE ENVIRONMENT, AND GENDER

TOGETHER THE ESSAYS IN THIS SECTION paint a picture of the complex relationships among gender, ecological health, and economic systems. The authors introduce us to the larger issues at stake in a conversation between ecofeminism and globalization.

1

Gender and the Environment

Mary Mellor

Introduction

Sociologist Mary Mellor provides a helpful overview of diverse ecofeminist approaches. She explores the claim that women are more responsive to environmental issues than men and traces various strands of ecofeminist thought, including analyses of global development's impact on women. While acknowledging diversity among ecofeminist thought, she concludes that all believe gender analysis is critical in addressing environmental problems.

T HE GENDER DIMENSION OF ENVIRONMENTAL ISSUES rests on two linked claims. The first is that women and men stand in a different relationship to their environment, that the environment is a gendered issue. The second is that women and men respond differently to environmental issues, in particular that women are more responsive to nature. Nature in this sense is more diffuse than the specific natural environment (the local ecosystem, the resource base of communities, and so on); it reflects a more holistic and active view of nature as a force. The term "environment" is used below to refer to the more limited meaning and "nature" to refer to the wider meaning. The claim that there is a gender dimension to environmental issues is initially the less contentious. It rests on the idea that, inasmuch as men and women have different life experiences, they have different environmental experiences. This idea becomes more problematic when extended to the assertion that environmental problems have more of an impact on women than on men. This in turn becomes linked to the second claim, that women are more responsive to environmental issues. Joni Seager, for example, has pointed out that women readily become active in campaigns about environmental issues and are overrepresented at the local level in formal environmental movements although underrepresented in the leadership of those movements.[1] There is also evidence that sexism and gender inequality in Green movements are reflected not only in the leadership profile but in Green ideologies.[2]

The claim that women are more responsive than men to environmental issues has been expressed in two ways. The first is based on women's different experience in a gendered society, arguing that women and nature are in a historically contingent relationship, that they have a socially constructed connection. The second sees the link as a more fundamental one: that women have an elemental affinity to the natural world based on biological or cultural sex differences.[3] In either case, raising the question of women's relationship to nature is very problematic for feminism, which has long sought to separate sex and gender. Nature in relation to women has tended to become entangled with embodiment, the perceived biological limitation and "weakness" of being female that has denied women political and social rights. As Simone de Beauvoir pointed out so forcefully, women appeared to be more prey to their biological destiny than men; they were locked in domestic and bodily immanence and could only gain freedom by rejecting and transcending their womanhood.[4] The case for reconnecting women with nature must therefore be a good one if all the gains of (some middle class, white) women are not to be lost.

This case has been made by the ecofeminist movement, which emerged contemporaneously in the middle of the 1970s in several different countries—France, Germany, the United States, Sicily, Japan, Venezuela, Australia, and

ECOFEMINISM SEEKS TO REESTABLISH RELATIONSHIP BETWEEN
WOMEN AND NATURE (AVENUE TO POWER?) WHEREAS
FEMINISTS WANT TO BREAK IT

Gender and the Environment   13

Finland,[5] although the French writer Françoise d'Eaubonne is credited with coining the name.[6] Ecofeminists argue that the reconnection of women with nature is necessary because the gendered nature of human society is directly related to the current pattern of ecological consequences. The risk for women that feminists see in opening up the woman-nature issue is justified by the need to confront the present scale of ecological destruction. Modern feminism in both its liberal and socialist forms has sought to rescue women from their association with nature and the body, although more recently the postmodern feminist position is more ambivalent. Ecofeminists do not see an equality or equal opportunities approach as the most progressive way forward. As Ynestra King, one of the founders of the ecofeminist movement in the United States, has argued, "What is the point in participating in a system that is destroying us all?"[7] If society is to go forward on a more sustainable path it may be necessary for feminists to retrace their steps and rethink the relationship between women, their environment, and nature more generally.[8]

From a feminist perspective, the most obvious way in which gender is linked to the environment is that most of the people who are in a position to affect environmental decision making are men and most of the people who are at the mercy of those decisions are women. However, this is not straightforward, as class and race crosscut gender in this analysis. Are women excluded from decision making and put at the mercy of environmental forces *as women* or because they are overrepresented among the poor, the exploited, and the colonized?[9] One of the key factors that has been identified in claiming both women's differential experience and awareness of environmental issues is the way in which women interact more closely with their local environment than do men. Where an environmental crisis occurs, women may be the first to notice foul water, obnoxious smells, or bodily ailments.

When Lois Gibbs began her protest over toxic waste at Love Canal, New York State, in 1978, neither she nor anyone else was aware that her housing estate and the children's school had been built over an abandoned toxic waste dump that was a mile long, fifteen yards wide, and up to forty feet deep. The first thing that Gibbs noticed was the unusual pattern of ill health within her family and among her friends and neighbors. It took considerable investigation and lobbying to find the cause of these problems and to get the residents relocated. Interestingly, it was just this pattern of awareness that Ellen Swallow had predicted a hundred years earlier. She had been the first woman to study at the Massachusetts Institute of Technology and could certainly claim to be the founder of the science of ecology. Swallow's interdisciplinary approach combined water chemistry, industrial chemistry, metallurgy, and mineralogy as well as expertise on food and nutrition. She established a laboratory to educate women scientifically at MIT, arguing that the home was the place where primary health and resources such as nutrition, water, sewage, and air could be monitored. Swallow's initiatives were not supported and her work became categorized as "domestic science." A man, the German Ernst Haeckel, is credited with inventing the subject of ecology in 1873. For Ellen Swallow, as has happened to women so many times, her contribution has been "hidden from history."[10]

It is, of course, true that where environmental problems affect local communities men are just as likely to be affected as women and children. In the United States, for example, there have been prolonged campaigns about environmental justice that have been based on class and race as well as gender. While ecofeminists have not disputed the importance of class and race, they have pointed not only to the way in which the gendered nature of society has put more women under environmental stress because they are disproportionately represented in low-income groups, but also to the fact that the gendered nature of Western society is directly related to the increased exploitation of the environment.[11] Although many of the early publications in ecofeminism concentrated on the experience of women in the West or North, in the 1980s the question of gender and the environment in the context of the globalization of western socioeconomic structures became increasingly central. Vandana Shiva has been highly influential in her analysis of the way in which male domination of modernizing economic systems, projected worldwide in the context of the development process, has undermined more sustainable ways of life. For Shiva, male-dominated destruction has been twofold: the global capitalist market system has systematically destroyed more sustainable ways of life that were associated with subsistence economic systems, and the inappropriate application of western science and technology has destroyed biological diversity and caused catastrophic ecological damage.[12]

The impact of globalized development on women has become increasingly important in the critique of development thinking, particularly in the 1970s and early 1980s, although the link between the impact on women and environmental consequences was only fully realized in the late 1980s. Both women's disadvantage and the environmental impact of the development process were exacerbated by western development agencies and workers who based their thinking on the gendered division of labor in industrial systems and failed to recognize the centrality of women in subsistence farming.[13] When common or family-owned land is privatized and turned over to cash cropping, women lose their right to land use. It is not only women's subsistence production that is at stake: women are universally the collectors of fuel and water and, as common open access land is lost, women have further and further to walk to secure these basic necessities.[14] As readily accessible and fertile land is lost for both subsistence farming and resource collection, women

REASON TO LOOK AT GENDER: WOMEN OFTEN BLAMED FOR ENV. DAMAGE, BUT REALLY DEEPER ECONOMIC ISSUES

are often forced onto more marginal (and ecologically fragile) ground. They have to cultivate thinning soils or collect green rather than dead wood. As a consequence, women, rather than the process of economic change that has forced them into this position, can sometimes be seen as the cause of environmental damage.

The first response to the failure to integrate women into the development process was a demand that women should be given the same economic opportunities as men, a campaign known as Women in Development (WID). However, as the ecological and social consequences of the development process became more apparent, a far more critical approach was taken. By the late 1980s, the campaign had shifted to a more critical stance under the influence of books and reports such as the Development Alternatives with Women for a New Era (DAWN) report and Maria Mies's and Vandana Shiva's work.[15] The WID campaign began to be replaced by a Women, Environment, and Development (WED) stance. This approach asked whether the development process was desirable any longer, certainly in its present western and male-dominated form. Central to the critique was the observation of the destructive effect on both women and the natural world.[16]

As in the case of the struggles over toxic waste in the United States, the question was raised as to how far the challenge to development and the global market system was based on specifically women's experience or that of colonized and peasant peoples generally. Should women's involvement in grassroots campaigns around the environment be seen as women's struggles or as peasant or communal struggles in which women played a part?[17] More contentiously, it could be argued that women's participation in these struggles was in some senses acting against their own interest, given the patriarchal nature of the traditional ways of life associated with peasant life and subsistence farming. The complexities of this situation can be shown by one of the best known of these grassroots campaigns, the Chipko movement. Based in the Himalayan hills, this movement gained international recognition for its direct action in hugging trees to prevent their logging by commercial firms. Vandana Shiva argues that, as the movement developed, it exposed gender differences in the approach to development within the local communities. Initially, men and women jointly opposed the transfer of the forest commons to commercial loggers and planters. However, village men and women had different ideas about the future use of the forest. While men wanted to create local commercial development by planting trees such as eucalyptus, women wanted to maintain and plant trees for fuel wood and fodder. It was at this point, Shiva argues, that the Chipko movement became "ecological *and* feminist."[18] (Italics in the original.)

The second area in which Shiva criticized male-dominated development was in the green revolution. Here scientists in western laboratories trying to meet world demand for food selectively bred heavy-cropping plant species without taking into account the local, social, and ecological conditions in which they would be used. In particular, the position of women farmers was not addressed. As a consequence, plants were introduced that were not suitable for local conditions, requiring large amounts of water, pesticide, and fertilizer. Ecological diversity was lost as local species were displaced and control of seed banks was maintained by commercial companies through the use of sterile hybrid plants.[19] As only the larger farmers could afford to use the new seeds, poorer farmers, including women, became impoverished, losing their land, or their access to land, to richer neighbors. Examples such as these appear to point to a systematic gender difference in relation to environmental issues, although these are crosscut by race and class. Women's and men's different social position means that they have different environmental needs and experience environmental problems differently. Even in poor communities women's disproportionate responsibility for family health and family subsistence differentiates their experience from that of men.

Although more recent ecofeminist thinking around the issues of gender and the environment has taken into account the experience of women in the so-called developing countries, most of the early ecofeminist writing was based on an analysis of gender divisions in Western society. There was a tendency to generalize from the experience of white, Western, middle-class women and their preoccupations or at least to speak of "women" in undifferentiated terms. However, there is an analysis at the heart of Western ecofeminism that can be seen as having a global applicability, since it focuses on the model of Western society that is being projected across the world in the process of globalization. This analysis directly links the gendered nature of Western society to the global ecological destruction that this model is creating.

The ecological destructiveness of the Western socioeconomic system has been seen by many ecofeminists as the result of the dualist nature of western society. Western society is seen as being divided in ways that prioritize one aspect of society through the denigration of its opposite or alternative. Scientific knowledge is valued over vernacular or popular knowledge; the public world of institutions and commerce is valued over the private world of domestic work and relations; abstract universalized thinking is valued over thinking linked to the particular and personal. For ecofeminists, these divisions are summed up in two crucial hierarchical dualisms: man, the masculine, is prioritized over woman, the feminine, and human society and culture are seen as superior to the world of nature. In these hierarchical relations, woman and nature are thrown into a contingent relationship as the despised and rejected by-products (or precursors) of "modernity."[20]

The origins of these dualisms are a matter of dispute among ecofeminists. For some, the divisions can be traced back to Greek society and the Aristotelian division between the public sphere of freedom and the private sphere of necessity and the Platonic division of the body and the soul. For some, it goes back even further, to the dawn of prehistory, when the benign world of the female earth-based goddess was overthrown by the destructive transcendent sky god.[21] For others, the division is historically closer, linked to the scientific and industrial revolutions that broke the traditional "organic" relationship between humanity and nature. Newtonian mechanics and the philosophical approach of people such as Francis Bacon saw the natural world as something inert and available for discovery and exploitation. All the earth's mysteries and resources would be opened up for "man." Merchant has argued that this approach spelled the "death of nature."[22]

The impact of the rejection of women and nature can be seen in the way both are devalued in commercial/industrial economic systems. Both have been treated as externalities in terms of economic accounting procedures and in the social construction of contemporary economic theory and practice.[23] The earth's resources have been seen as either free (air and oceans) or only worth the cost of extraction or the compensation paid to those who own or occupy the relevant areas. The prices of primary products are determined by the level of wages that can be set in disadvantaged countries and the vagaries of the casino financial market in these products. Long-term costs or responsibility for polluted or depleted resources do not appear on the commercial accounts of companies benefiting from natural resources or primary production. Equally, women's work has been devalued.[24] Most of women's work across the globe is either unpaid or paid at a low rate. Ecofeminists argue that this is because women's work is associated with the bodily process of life, from child care and hygiene to health provision and basic food production. In their common marginalization, women and nature appear to have been thrown into at least a contingent relation. Does this mean that women are in an epistemologically privileged position in terms of environmental questions? Are women more responsive to nature?

As has been pointed out, there are two broad approaches to this question. One stresses women's socially constructed relationship to the natural world, while the other sees a much deeper affinity. Social or socialist ecofeminists see women's closer relation to the natural world as socially constructed. Any superior knowledge women may have about the environment or the natural world stems from their social position. Affinity ecofeminists see women as closer to the natural world through their embodiment as women and/or mothers or as the representatives of a feminine cosmic force. Often, however, the division is one of rhetoric. The U.S. writer, Susan Griffin, for example, can be seen as taking a deeply essentialist position in her text *Woman and Nature*, a prose-poetic rendering of the dualist voices of "scientific man" and "natural woman." It is, however, clear from her later writings that she takes a social constructionist position on gender divisions.[25]

One of the reasons for ecofeminism's association with an essentialist radical feminism is its emergence alongside the cultural feminist radicalization of the feminist movement, particularly in the United States, and the deepening of the green movement after Arne Naess had pointed to the difference between a shallow and a deep ecology.[26] In North America there are particularly strong links between ecofeminism and cultural feminism and the feminist spirituality movement. Within the feminist spirituality movement, the gendered divisions of modern society are seen as representing a cosmic division between the forces of the feminine and the masculine, the god and the goddess. This influence, reflected in two well-known anthologies (Plant in 1989 and Diamond and Orenstein in 1990),[27] led to accusations that ecofeminism was irrational and reactionary in terms of modernist feminist aims (Biehl in 1991 and Evans in 1993).[28] However, even in these texts, the work of social constructionist ecofeminists such as Ynestra King, whose roots lie in anarchism, or Carolyn Merchant and Rosemary Radford Ruether, who adopt a basically socialist position, is also represented. All three see merit in the cultural feminist arguments, particularly in relation to the analysis of patriarchy.

There are also very few affinity ecofeminists who take an ultimately essentialist line on gender. Most culturally based ecofeminist writers do not see a cosmic and universal, unbridgeable difference between men and women. However, many ecofeminists see women as having an affinity with the natural world that men do not have. Petra Kelly, the late German Green activist, argued that a woman could "go back to her womb, her roots, her natural rhythms, her inner search for harmony and peace, while men, most of them anyway, are continually bound in their power struggle, the exploitation of nature, and military ego trips."[29] What appears as a biologically determinist argument is muted by the phrase "most of them anyway" in relation to men. While women are seen as biologically connected to the natural world, men are not *biologically* disconnected. A similar qualification occurs in the work of one of the most vociferous exponents of affinity ecofeminism, Andrée Collard, who asserts in her book, *Rape of the Wild*, that "the identity and destiny of woman and nature are merged."[30] In the book, Collard argues that women are linked to the natural world through their ability to give birth and nurture (even if they have never had children). Men are not inherently destructive, however: it is patriarchy, not men per se, that is the enemy of nature. Collard does refer to men as if they and patriarchy were one. At the end of the book, she praises the political action of men who are ecologically sensitive. It seems that women, whether they are mothers or not, are condemned to their affinity with nature, whereas men can choose. Ynestra King argues that women also have a choice. Given that society has been socially constructed in such a way that women and the natural world are forced into an alliance, women can choose whether to reject that association or to maintain it for political reasons. Women can "*consciously choose* not to sever the woman-nature connection by joining male culture."[31] (Italics in the original.)

Most ecofeminists, whether affinity or social constructionist, take what approximates to a standpoint perspective. Women—having been biologically, cosmically, or socially placed in a subordinated position alongside a devalued natural world within Western/patriarchal, dualist, socioeconomic structures —are better placed to see the way in which social relations have an adverse impact on the natural world than are men from their superordinate position.[32] There is, however, ambivalence in ecofeminist writings about whether women will spontaneously see the woman-nature relation merely through their subordinate position, or whether the position of women is to be the starting point for an analytical framework and activist campaigning. For both affinity and social constructionist ecofeminists, the basic argument is not that women are essentially or biologically closer to nature, but that (superordinate) men are distanced from their natural environment in dualist structures. In particular, they are distanced from the ecological consequences of their actions and the biological needs and limitations of their embodied existence. The physical burden of these ecological consequences and the meeting of biological needs (physical comfort, hygiene, food and shelter, care in maturation and infirmity, and so on) are borne by others. Women who adopt superordinate positions can also lose touch with the natural roots of human existence, but it is harder for them to cast aside domestic and other caring responsibilities. Equally, men in subordinate positions may bear the burden of embodiment or suffer ecological consequences, particularly in industrial ill-health, but the sexual division of labor within households and communities still leaves women with the major responsibilities for human embodiment.

Ecofeminists argue that a Green perspective is not adequate if it does not see the way in which the gendering of society produces adverse ecological consequences. They differ in their explanations of how dualist structures produce a gender and ecological "blindness." Those whose discipline base is in philosophy tend to point to the "logic of domination" inherent in Western philosophical systems which produces dualist structures of thought, which is generally traced back to the Greeks. Epistemological privilege here rests with those women (and men) who are able to break out of that framework: they form the epistemological "bridge" between nature and culture. Those with a theological or spiritual base tend to see dualist structures as representing a more fundamental battle between cultural forces in religious structures. The struggle is more universal and cosmological, yet again men seem to be able to "jump ship" from patriarchal ways of thinking and embrace more earth-centered spiritualities. For those whose discipline base is in the social sciences, more materialist explanations are offered. Stress is put on women's work in society, particularly around human embodiment. Women are seen as being placed structurally closer to the natural functions of human existence in a way that allows dominant males to escape to a transcendent public world. Such a position puts less stress on the naturalness and spontaneity of women's identification with the natural world and much more on the structural and material relation of women to nature as the starting point for critical analysis.[33]

Although ecofeminists may differ in emphasis and analysis, they share the viewpoint that a gender analysis is essential if ecological problems are to be addressed. Most would extend this analysis to race and some to class. Gender inequality is seen as producing male-dominated social structures, which become detached from their environmental context and therefore lose awareness of the impact of human activity on the natural world. Ecofeminists see the environmental consequences of "modernizing" global structures as being disproportionately inflicted on women, indigenous communities, marginalized and exploited peoples, and on the natural world and its nonhuman inhabitants. If women have epistemological privilege, it is as part of a matrix of subordinated structures whose subordination creates the illusion of Western, male-dominated modernity and progress based on economic and ecological exploitation.

Notes

1. Joni Seager, *Earth Follies* (London: Earthscan, 1993).

2. Mary Mellor, "Green Politics: Ecofeminist, Ecofeminine or Ecomasculine?" *Environmental Politics* 1, 2 (1992): 229–51; Ariel Salleh, "The Ecofeminism/Deep Ecology Debates: A Reply to Patriarchal Reason," *Environmental Ethics* 14 (fall 1992): 195–216.

3. Mary Mellor, *Breaking the Boundaries: Towards a Feminist Green Socialism* (London: Virago, 1992); Mellor, "The Politics of Woman and Nature: Affinity, Contingency or Material Relation?" *Journal of Political Ideologies* 1 (2): 147–64.

4. Simone de Beauvoir, *The Second Sex* (London: Jonathan Cape, 1968).

5. Valerie Kuletz, "Interview with Barbara Holland-Kuntz," *Capitalism, Nature, Socialism* 3 (2): 21–25; Ariel Salleh, "Discussion: Eco-socialism/Eco-feminism," *Capitalism, Nature, Socialism* 2 (1): 129–40.

6. Françoise d'Eaubonne, *Le Feminisme ou la Mort* (Paris: Pierre Horny, 1974).

7. Ynestra King, "Healing the Wounds: Feminism, Ecology and Nature/Culture Dualism," in *Reweaving the World*, ed. Irene Diamond and Gloria Feman Orenstein (San Francisco: Sierra Club Books, 1990), 106.

8. Mary Mellor, *Feminism and Ecology* (Cambridge: Polity Press, 1997).

9. Maria Mies, Veronika Bennholdt-Thompson, and Claudia von Werlhof, *Women: The Last Colony* (London: Zed, 1988).

10. Patricia Hynes, "Ellen Swallow, Lois Gibbs and Rachel Carson: Catalysts of the American Environmental Movement," *Women's Studies International Forum* 8, 4 (1985): 291–98; Celeste Krauss, "Blue-Collar Women and Toxic-Waste Protests: The Process of Politicization," in *Toxic Struggles*, ed. Richard Hofrichter (Philadelphia: New Society Publishers, 1993); Robert Clarke, *Ellen Swallow: The Woman Who Founded Ecology* (Chicago: Follet, 1973); Anna Bramwell, *Ecology in the Twentieth Century* (London: Yale University Press, 1989); Sheila Rowbotham, *Hidden from History* (London: Pluto, 1973).

11. Mellor, *Breaking the Boundaries*; Ariel Salleh, "Nature, Woman, Labour, Capital: Living the Deepest Contradiction," in *Is Capitalism Sustainable? Political Economy and the Politics of Ecology*, ed. Martin O'Connor (New York: Guilford Press, 1994).

12. Vandana Shiva, *Staying Alive* (London: Zed, 1989).

13. Shiva, *Staying Alive.*

14. Gita Sen and Caren Grown, *Development Crises and Alternative Visions* (New York: Monthly Review Press, 1987).

15. Sen and Grown, *Development Crises*; Maria Mies, *Patriarchy and Accumulation on a World Scale* (London: Zed, 1986); Shiva, *Staying Alive.*

16. Rosi Braidotti, Ewa Charkiewicz, Sabine Hausler, and Saskia Wieringa, *Women, the Environment and Sustainable Development* (London: Zed, 1994).

17. Bina Agarwal, "The Gender and Environment Debate: Lessons from India," *Feminist Studies* 18, 1 (1992): 119–58; Cecile Jackson, "Radical Environmental Myths: A Gender Perspective," *New Left Review* 210 (March/April 1995): 124–40.

18. Shiva, *Staying Alive*, 76.

19. Vandana Shiva, "The Seed and the Earth: Biotechnology and the Colonisation of Regeneration," in *Close to Home*, ed. Vandana Shiva (Philadelphia: New Society Publishers, 1994).

20. Val Plumwood, *Feminism and the Mastery of Nature* (London: Routledge, 1993); Karen Warren, ed., *Ecological Feminism* (London: Routledge, 1994); King, "Healing the Wounds," in *Reweaving the World.*

21. Rosemary Radford Ruether, *New Woman, New Earth* (New York: The Seabury Press, 1975); Plumwood, *Feminism and Mastery*; Riane Eisler, *The Chalice and the Blade* (London: Unwin, 1990).

22. Carolyn Merchant, *The Death of Nature* (New York: Harper and Row, 1983).

23. Marilyn Waring, *If Women Counted* (London: Macmillan, 1989); Mellor, "Green Politics," *Environmental Politics.*

24. Sheila Lewenhak, *The Revaluation of Women's Work* (London: Earthscan, 1992).

25. Susan Griffin, *Woman and Nature: The Roaring Inside Her* (New York: Harper and Row, 1978); Susan Griffin, "Split Culture," in *Healing the Wounds*, ed. Judith Plant (London: Green Print, 1989).

26. Arne Naess, "The Shallow and the Deep, Long Range Ecology Movement," *Inquiry* 16, 4–5 (1972): 95–99.

27. Judith Plant, ed., *Healing the Wounds* (London: Green Print, 1989); Irene Diamond and Gloria Feman Orenstein, *Reweaving the World* (San Francisco: Sierra Club Books, 1990).

28. Janet Biehl, *Finding Our Way: Rethinking Ecofeminist Politics* (Montreal: Black Rose Books, 1991); Judy Evans, "Ecofeminism and the Politics of the Gendered Self," in *The Politics of Nature*, ed. Andrew Dobson and Paul Lucardie (London: Routledge, 1993).

29. Petra Kelly, *Fighting for Hope* (London: Chatto and Windus, 1984), 104.

30. Andrée Collard and Joyce Contrucci, *Rape of the Wild* (London: The Women's Press, 1988), 137.

31. King, "Toward an Ecological Feminism," *Healing the Wounds*, 23.

32. Nancy Hartsock, *Money, Sex and Power* (Boston: Northeastern University Press, 1983).

33. Mellor, "Green Politics," *Environmental Politics*; Salleh, "The Ecofeminism/Deep Ecology Debates," *Environmental Ethics*; Plumwood, *Feminism and the Mastery of Nature*; Warren, *Ecological Feminism*; Charlene Spretnak, *States of Grace* (New York: Harper Collins, 1991); Salleh, "Nature, Woman, Labour, Capital"; Mellor, *Feminism and Ecology*.

2

Can Ecofeminism Withstand Corporate Globalization?

Heather Eaton

Introduction

Heather Eaton, an ecofeminist theologian, reflects on how corporate globalization is reordering the world and where ecofeminism might intersect. Heather uncovers several faces of an anonymous, yet powerful, corporate rule. She examines what kinds of theological methods could enable ecofeminism to be of significance in the face of global corporate structures and reveals the need to be aware of the larger, especially economic, forces of the world.

A revised version of Heather Eaton, "Ecofeminism and Globalization," *Feminist Theology* 24 (May 2000). Copyright © 2000 by Sheffield Academic Press. Reprinted by permission.

THERE IS A NEED TO BRING THE GLOBAL ECONOMIC AGENDA into the center of ecological and feminist religious reflections if ecofeminist analyses are to be pertinent to the global economic ghoul that is shaping many levels of current reality. One of the questions is whether ecofeminist analyses, and in particular ecofeminist religious viewpoints, are effective in the face of economic globalization, or whether they are essentially just inspirational. In a desire to evaluate the power and liberatory potential of ecofeminist discourses in light of globalization, and in particular global corporate rule, I began this inquiry. A further desire was to expand the capabilities of ecofeminist liberation theologies to confront globalization. This essay is a personal reflection on the process of attempting these goals in the context of ecofeminism, theology, and globalization. In brief, the results have been a disturbing realization of the power of corporate rule, and the fallibility of ecofeminism.

In an initial approach to the relationship between ecofeminism and globalization, I thought of potentially useful goals for the reflection: 1) to challenge the theoretical framework of religious ecofeminism and move beyond the cultural, ideological, and conceptual connections between women and nature to the concreteness of ecological and related social issues, such as deforestation, drought, pollution, biodiversity losses, militarization, and socioeconomic impoverishment; 2) to support the need for religious critiques of the dominant global systems of economic profit that function through oppression of those who benefit least and leave ecological and social ruin in their wake; 3) to suggest paths of liberation from the vantage points of ecofeminist liberation theologies, North/South experiences, and radical religious movements.

As I began to examine ecofeminist religious discourses in light of globalization and these categories, I became overwhelmed by the governing reality of global corporate rule. This was followed by a wave of anxiety and an awareness of vast discrepancies between corporate rule and ecofeminist religious discourses. Having worked for many years as an ecofeminist liberation theologian with an acute sense of concern for what is happening in and to the world, indeed to all life on earth, the incongruity was startling.

To become informed about globalization is to find ambiguous definitions and hydra-headed characteristics. Globalization has many expressions. Two basic usages are the following. First, globalization refers to the shrinking of space and the vast intersecting of culture, technologies, religions, communications, and ideas–the "global village." Second, globalization, or the global economy, is equated with external market liberalization and a reliance on the equitability of market forces. Often these two usages are intertwined and confused. For this essay it is the ideology of economic globalization that is most important and troubling, and it is embedded within the above two expressions.

Ideologically, economic globalization appeals to an ideal of adventure, entrepreneurship, and superiority. "Gateways to the World," "Go Global," "Track Global Competition," "Spread Global Wings," "Crossing International Borders," and "Becoming Master of One's Domain"—these slogans invite expressions about global prospects for business. Any mainstream business magazine is filled with these and similar expressions. They create a perception that the world is one homogenous reality ripe with economic potential that can be plucked by the adventurous. One can rise above context and place to thrive in a virtual globe of riches with little or no constraints. Implicit is that business is the greatest possible model of life, far superior to cultures, nation-states, bioregions, and so on. There is no talk of differentiated and distinct cultures of people, of ethnicity or gender, of animals or land. There is no conversation about national or international regulations, or even that there are genuine limitations to this frontier of capital exchange. In the ideological jabber there is little discussion of the complex issues around government policies on political, economic, and social choices with respect to the terms upon which any given country could or should relate to the globalizing of their economies. There is virtually no discussion of the uneven flow of capital and information. This "globe" of which they speak is an utter abstraction with no accountability to anything but the hegemonic economic agenda.

The consequences of this seductive rhetoric are never mentioned. What about the mining corporations in Latin America that use cyanide in the water to separate minerals to produce gold for jewelry? The water table has become saturated with cyanide from which animals, plants, and people drink. Further unacknowledged repercussions are the relocation of workers and the resulting fragmentation of community structures; the increasing sexual exploitation of women and children by men as the social fabric weakens; the loss of home, land, and livelihood because land is now owned by a multinational corporation; and the escalation of toxins in air, land, food, and in most life forms. The lives of many, especially poor women, are marked by an increase in work and a decrease in health. As a result of some globalization initiatives, there is a deterioration in educational and health systems, a rise in infant mortality, and a decline in democratic pluralism. In addition, communities and countries are coerced into export-dependent economies.[1] Social norms and fabrics shred under the force of anonymous corporate "restructuring" in the name of globalization. John Jordan comments that "transnationals are affecting democracy, work, communities, culture and the biosphere."[2]

The growing numbers of poor people bear the direct and immediate costs of this dysfunctional system, yet they are often stripped of decision-making power as the transfer of regional resources to larger institutions increases. These "systems" are frequently inaccessible to local people and oblivious to their needs. Further, many people are kept in a state of confusion by corporate media regarding the causes of their distresses. Those who resist are held hostage by layers of a system that is not accountable for the consequences of its actions—not to people, land, or animals—only to the GNP. But what does the GNP measure? Feminist economist Marilyn Waring showed that much of life's work, caring for children or the aged, cottage industries, subsistence farming, and women's work in particular, are not calculated into the GNP. It is even more skewed, and repulsive, when social and ecological disasters such as the Gulf War or oil spills actually register as a gain to the GNP.[3]

From this vantage point, globalization indicates an erosion of democracy as power shifts from governments, which are supposed to act for the common good, to a handful of corporations whose only goal is short-term economic gain. The systemic forces nurturing the growth and dominance of global corporations are at the heart of the current human-earth dilemma. Since 1994 and the formation of the World Trade Organization (WTO), governments have been in a permanent hostage situation to the global economic system. We are living in a political era of corporate rule that is determining government policies. Economist David Korten said that the world is now ruled by a global financial casino, and that democracy is for sale to the highest bidder,[4] because the WTO is so closely aligned with corporate rule. They have granted over five hundred corporations security-clear trade credentials, and in many parts of the world, such as Japan and Europe, relationships between corporations and governments are firmly institutionalized. Thus their power through the WTO is enormous.[5]

This view of globalization forces the question of the need for theology to be *attending to the world*, reading the signs of the time, and being something of significance in this global reality. What world are theologians attending to? Much of theology is inward-looking and yesteryear-referenced. There are endless conversations/disagreements about texts, dogmas, beliefs, and conventions. I have often criticized theology for its blatant failure to attend to, or even take interest in, the concrete ordering of the world—that of power, control, access to resources, and overwhelming suffering. It is disturbing that mainstream theology continues to neglect the pervasive devaluing of women, the marginalized, and the natural world. This autistic theology continues to be the hegemonic form. Liberation theologies are the exception. I look to the ecofeminist liberation approach to be the only, even potentially adequate, voice that is attending to this world. In comparing these religious responses to the globalization reality, it is evident that much of theological discourse is inadequate. I look at ecofeminist theology—including my own hard work in ecofeminist liberation theology for over a decade—and see little that can challenge corporate rule. What is the relationship between theologies of liberation and the globalization meta-narrative and concrete reordering of the world?

Liberation theologies evoke elaborate theories about how to begin critical analysis of our social relations as the only way to achieve nondualistic, relational, liberatory and, as Beverly Harrison remarks, intrinsically historical and time-bound theologies.[6] The hallmark of liberation theologies is, in principle at least, the dedication to reality, which is concrete and historical-bound. An analysis is theological if, *and only if,* it unveils or envisions our lives as a concrete part of the interconnected web of all our social relations, including our relations to God.[7] I would add our relations to the natural world, which means a consciousness of both our ecosocial locations,[8] as well as an examination of our worldview from a cosmological perspective, as advocated by Thomas Berry[9] and Rosemary Radford Ruether,[10] for example.

There are numerous promising theological voices engaged in a reformulation of theology. For example, ecofeminist liberation theologies have expanded and deepened, and they now have several interlocutors, including social, feminist, political, cultural, and postcolonial theories, theories of emancipation, critical theory, and a variety of postmodern analyses. Ecofeminist theologies take their cues from feminist and ecofeminist theories, as well as from theological traditions. The dialogue partners are numerous.

Ecofeminist theologies call for reinterpretations of the *foundations* of theology, as do feminist and ecological theologies. There is innovative work occurring in the reworking of doctrine, biblical motifs, rituals and, to some extent, theological method. There is a fluidity of images and much creativity. Yet when I look at the great bestsellers in feminist and ecological theology of late, there seems to be a chasm between theory and transformative praxis. Which voices can address globalization?[11] We have myriad theories and ideals of relationality, respect for diversity and ethics of mutuality. We can stand for empowerment and earth ethics. We can name the "Changing of the Gods" and call forth "She Who Is," but still, where does all of this intersect with globalization? Are we engaging in the same disconnected theology in a more subtle, consoling form? Marked by the wavering critical spirit of the Enlightenment, carrying a variety of postmodern analytic tools, and perhaps drained from trying to change the world, is there a subtle abdication—resignation—to theoretically sophisticated yet anaemic discourses? Are we drinking father's milk through female authors, such as muse Catherine Keller suggests?[12] The more ecofeminist theologies are detached from globalization—detached from knowing the evidence and detached from the "frail global networks of accountability"—the greater the chance will be that we are promoting liberal, albeit graceful, theologies with little or no political responsibility.[13]

Meanwhile the disintegration of the earth accelerates. Water is becoming an international market-commodity, not a basic need or right. The militarization of some countries increases to ensure corporate power. The Peace Research

Institute in Oslo found that most of the civil wars of the past decade have taken place in countries with high poverty levels, little fresh water, land degradation, high external debts, and a history of vigorous International Monetary Fund (IMF) intervention. They concluded that most of these difficult conditions were heightened, if not caused, by the World Bank, the WTO, and the IMF.[14]

Michael Jordan was paid $20 million for promoting NIKE shoes, which is far greater than the total annual payroll of all of the employees in the Indonesian factories that make the shoes. The twelve- and thirteen-year-old girls, with no protection of any kind, are paid fifteen cents an hour to make a shoe, which is bought in Indonesia for $5.60 and sold for between $75 and $135.[15] Michael Eisner, former chairman of Walt Disney Corporation, was given a personal executive package of $203 million—for the illusions of the wonderworld of Disneyland.

Economic globalization is predicated upon the illusion of an ideal world in which well-being is measured by the accumulation of things to ease the burdens of life's harsh conditions. But often the wheat and the chaff are indistinguishable. For example, the availability of health care throughout the world is necessary. As governments are less willing to assure this and pass the responsibility to multinationals, access to and quality of health care are inconsistent. Because it is more profitable, there is deliberate withholding of generic drugs by multinational pharmaceutical corporations, such as in the case of the AIDS epidemic in Africa. We see the same refusal to provide generic drugs in Canada, so as not to disrupt corporate profits. The need for reliable food sources is also crucial, but under the biotechnology banner toting the "feed the world" slogan, multinationals undermine local sustainable farming communities by taking their land to export agrobusiness cash crops.[16] Native seeds are displaced with sterilized hybrids, and the inevitable results are increased pesticide usage, aridity, imported diseases, and soaring costs. There is much ambivalence about this new global economy, says Vandana Shiva. It is wrong to be so smug about its promises, especially due to the negative impact of globalization on specific regions and on the lives of the majority of people who remain rooted in their local contexts.[17]

The contested "terminator technology," a genetic modification technology that sterilizes seeds such that they need to be repurchased for every sowing, has now been accepted for commercialization by the United States Department of Agriculture (USDA) as of August 1, 2001. There has been overwhelming opposition to and condemnation of this technology, as it blatantly puts private profits above public good and the rights and livelihoods of farmers everywhere who need farm-saved seeds. International pressure came from all quarters, including from members within the United States Biotechnology Advisory Committee, who urged the USDA to abandon these patents and stop further research on genetic seed sterilization. Still the USDA licensed it.

What about the few molecules of PCB released in Big Spring, Texas, that travelled though several countries and ended up in seals, polar bears, and the breast milk of the Inuktitut peoples in the northern isolated island of Broughton? This is not a unique case, but rather an example of the fact that there is no safe, uncontaminated place. What about the charming fact that in six months of breast-feeding, a baby in Europe or North America gets the maximum lifetime recommended dose of dioxin and five times the allowable daily level of PCBs set by international standards for a 150-pound adult?[18]

What are the causes of the global crisis? Not globalization, according to the corporations. Rather it is government restraint of markets: "give trade not aid" is the advice of the corporate elite to the IMF. What actually brings wealth is economic liberty, nestled within a stable cultural and legal framework: the refrain from any *World Trade Magazine*. Trade restrictions are causing decreases in GDP. They say that we (whoever that is) must create free economic markets and more free trade zones. The results will be greater GNP and GDP. Of course "we" all know that this really means more wealth for the elite. For the majority there is an increase in slavery working conditions, child labor, no environmental regulations, no international law, no union support, "encouraged" sterilization of women, increased illness, and life amidst the toxic ruins of the earth. This has been documented over and over for all of the free-trade zones—the epitome of freedom. If the General Agreement on Trade in Services (GATS) becomes the global charter, then national laws on such matters as economic justice, worker health and safety, education, social security, and ecological regulations can be struck down by the WTO if they are determined to be barriers to trade.

To pause briefly on the issue of water reveals the extent of the problem of globalization. The World Bank and the IMF want water privatization. They promote monopolies and have refused loans to refinance water services unless governments privatize water systems. There is a close alliance between governments, the World Bank, and water corporations (for example, Vivendi SA, Suez Lyonnaise des Eaux). It is water shortages that are the driving force for the privatization of water trade. Although the language is to make water accessible and competitive, this is not the result. In most cases of water privatization, there are increased shortages and increased contamination. International trade agreements are the tools corporations are using, such as Chapter 11 of the North American Free Trade Agreement (NAFTA). It goes like this. Foreign companies can sue governments on laws, rules, or regulations, at any level, if these rules are impinging on the right to make a profit—based not on monies actually lost but on future profits. For example, Methanex, a Canadian company with U.S. subsidiaries, is suing the U.S. government for $970 million. Methanex produces a gasoline additive called MBTE (methyl tertiary butyl ether), which is being banned in California because it is toxic and has leached into the water table in thousands of places. It is linked to cancers and possible neurological and dermatological problems, says California. Methanex says there is no conclusive proof of a health hazard (it only takes one dissenting scientist to constitute no proof), although eleven states are in the process of banning it. Instead of being concerned that they could be sued for the closing of municipal wells in Santa Monica and the sickness caused, Methanex is suing the federal government for substantial interference, or expropriation. The corporation is the victim, claiming damages for present and future financial losses. Corporations have a set of legal resources at their disposal that have never existed before. The NAFTA negotiations and tribunals take place in private and at the federal level, meaning that not only are citizens' voices excluded, but California, in this case, may have no voice in the debate. NAFTA gives the federal government power to override state actions that are contrary to the trade agreement. There is no appeal.[19]

The corporate world tries to control who gets to know what. Countless stories demonstrate that corporate power supported by government—militia, media, and/or legal representation—can prevent the underside narrative from becoming public. The tragic execution of Ken Saro-Wiwa in Nigeria—because he exposed the ecological and social devastation caused by Shell Oil and the government—reminds us that the benevolent corporate image is a facade.[20]

Corporate "green washing" is big business. Major corporations are now "environmentally friendly," engaging in the great oxymoron of sustainable development and producing green products. Car makers, aerosol manufacturers, oil producers, the nuclear industry, and the forestry business—some of the worst pollution producers—are all suddenly ecofriendly, or so say their glossy green ads. Corporations in these businesses have changed their images. Some have created green business networks and green front organizations to act as their representatives. Some have had the audacity to create Non-Governmental Organizations (NGOs) in order to receive funding and be considered legitimate at environmental gatherings of the "alternative" NGOs. Corporate environmentalism has become the true ecological pioneer, at least according to Bruce Harrison, author of *Going Green: How to Communicate Your Company's Environmental Commitment*. At least he had the honesty to state that for business "getting on the green is not easy."[21] Welcome to the land of green babble, where conventional environmentalism has been replaced by envirocomm—environmental communication.[22]

The corporate world has colonized everywhere: from television to classrooms, painting themselves green, supporting women's initiatives, universalizing the consumer, and commercializing youth. Multinational corporations are involved in energy, biotechnology, agriculture, food-processing, manufacturing and retail, communications, transportation, media, health, and education. In Canada the corporate world is restructuring education, supporting certain programs and eliminating others—liberal arts for example. They give millions to universities to develop biotech programs. The education system is capitulating because government funds are limited. The results of the loss of the reflective disciplines are becoming apparent. Students may have acquired some data, but few are skilled in critical thinking and discerning the difference between data, information, knowledge, and wisdom.

From 1990 to 1995 there was $4.5 trillion in corporate profit, and the U.S. government gave out over $125 billion in corporate welfare—tax cuts, exemptions, credits—not including the relaxing of environmental regulations, the benefits from the WTO, intellectual property rights, and so on. On the meta scale, many corporations have larger economies than countries. Mitsubishi and General Motors have larger economies than Indonesia, Denmark, and Thailand. With the addition of Walmart, these and numerous other corporations have larger economies than Hong Kong, Saudi Arabia, Israel, Greece, Iran, Chile, and Egypt. Out of the top one hundred economies of the world, forty-eight are countries, fifty-two are corporations.[23]

How can ecofeminist theology respond? What theoretical frameworks are adequate to match the corporate narrative? There is a great need for thinking at the systemic level–whole-systems thinking. Do we engage in the foundational thinking? Or is this another form of hegemonic oppressive discourse? Corporate rule is a meta-narrative. Feminists deconstruct meta-narratives. Postmodern, poststructuralist narratives argue that meta-narratives are archaic. Certainly this emphasis is necessary for the emancipation and appreciation of human/cultural distinctions and the decentering of hegemonic interpretations of reality, but globalization is now the meta-narrative.

Theology has many dialogue partners and attends to diverse theories. What theories are adequate for theology these days? *Which theology and for whom* is the obvious reply. Still, how can the nature of theology be reconceptualized in light of globalization? The goal is not to reflect on the occurrence of globalization from an observational podium, but to confront and transform through a theological voice of resistance with a vision of a viable alternative future.

How do we use theory in ecofeminist theology, and what are the limits of theory? Rebecca Chopp addresses several of the intricate pieces of this question in terms of the use of feminist theory in theology. The emphases on methodology, and in particular epistemology, have strengthened feminist discourses, but there are losses, such as a lack of attending to the material world, or to the grand, seemingly utopian, goals of ecofeminism. Chopp contends that rather than avoiding the global situation or the utopian visions of feminism, feminist theology "might think even harder about the use of theory."[24]

The theory/praxis dialectic is an essential preoccupation of feminist theories committed to concrete liberatory changes in the situation and lives of women. I am drawn to Rebecca Chopp's notion of pragmatism: strategies of truth through either culturally situated communities or complex traditions that empower human flourishing.[25] Theories are combined for ultimate aims, she suggests, and are resistant to any meta-theoretical framework (what is contained in "ultimate aims" is surely a question). For example, if change is only possible at the local level, or further still, if the globalization system can only be dismantled at the local level, there remains an uncompromising goal to theorize how to end this specific aspect of corporate rule and globalization. Theories become the strategy for change. Unfortunately the theories functioning within movements for "change" are often clandestine. Perhaps the next best idea is to think in terms of the concrete consequences of theorizing and theologizing, albeit a difficult task that requires observation time—an endangered "commodity." Perhaps, as suggests Mary Grey, it is time to hear from those acting and reflecting directly out of their own struggles, and not from those of us trying to include our voices.[26]

Perhaps it is best to go the route of poststructuralism: to attend to the particular, the unique, giving priority to differentiation, to specific contexts and to the subject, subjectivity, and the local. Julia Kristeva, and to some extent Emmanuel Levinas, would suggest that foundational thinking is totalitarian, "all global problematics are archaic . . . [we] should not formulate global problematics because this is part of the totalitarian and totalizing conception of history."[27] There is strong feminist work in support of postmodern, poststructuralist versions of theorizing and theologizing.

Yet what are the dangers of this level of particularization and detachment from the whole? Beverly Harrison sees the need to address "particular theories in particular locations" and feels this should be the primary focus of feminists.[28] However, Catherine Keller has observed that the emphasis on difference and particularity, and the deconstructive postmodern discourse in general, while necessary for the emancipation and appreciation of human distinctions and the decentering of hegemonic interpretations of reality, are hindrances to the development of a comprehensive ecological awareness.[29] Yet the postmodern aversion to addressing the systemic sufferings of the world may camouflage an inability for poststructuralist thought to engage adequately with a level of systemic analysis oriented towards globalization and resisting its effects.

What has happened to the axiom of Marx, that of moving from interpreting the world to changing it? Critical theory is an ally to this axiom and potentially could assist liberation theologies.[30] The starting point of critical theory is the oppression and suffering of a particular society, and it aims to expose the structure of relations causal to the distress. Critical theory can address the operations of knowledge in the deliberation of beliefs, activities, illusions, and social constructions of the community.

There are still other questions. What language can be used to address globalization? What language is liberatory? How can anyone speak of the "waters of life" in the face of the global water crisis? If, as any mainstream magazine claims, love can refer to a car, purity to a detergent, gin to an infinite value, or retirement planning to a revolution, then language is governed by consumerism and the ideology of globalization. Religious images have been taken over. Dorothee Sölle says that religious language has been destroyed and corrupted. In a culture that expects all of us to be informed hourly about cat food and hair spray, life is insignificant. She says that what cannot be sold is worthless, and our ability to perceive has been disturbed and our feeling for reality trivialized. Sölle writes, "and the sacredness of life for which I am here trying to plead is consistently and pitilessly destroyed in the rituals of consumerism."[31]

In the "developed" countries, the amount of clothing, shoes, gadgets, toys, cars, cell phones, computers, techno-upgrades, renovations, waste of all kinds . . . is based on ideological messages that these make the world a better place and fulfill human dreams and aspirations. The powerful seduction of this ideology of wonderworld is, in fact, creating a waste world, says Thomas Berry.[32] The illusion of wonderworld is weakening, but it has sunk deep into Western intelligence and spirituality; it remains the prototype of virtually every profession. It is a mesmerizing, hypnotizing fantasy that dazzles us, like Siren's song, to our demise.

How can we say what we want and expect from life? Corporations manipulate the cultural and religious symbols in which our individual and communal identities and values are anchored. We are promised salvation by this or that gadget, car, item of clothing, food, house, sports . . . whatever. They feed on genuine human needs and desires, such as to be known and accepted, comforted and cared for. Corporate consumerism is defining who we are, and what it means to be human.

Shame is a revolutionary language, says Marx. We need to feel shame at how religious language has been trivialized. It is sad when an advertisement says that liberation is a priority status on airplanes, meaning that we can choose between elite, super elite, or executive class, all on the same plane a few feet from one another. It is tragic, indeed an ultimate obscenity, that the rich sit in comfortable living rooms watching people starve, be victimized by violence, or have their homelands vanish due to a natural disaster that is likely caused by the culprits of global warming.[33]

One of the advertising claims of Mitsubishi is that it "redefines how you see the world." What language will allow us to see the world through religious eyes—not utopian or romantic, but with a dedication to reality? The reality that needs some dedication is globalization.

Ecofeminism, a voice—a language—of resistance and vision, comes from many sources, two of which are activism and theory. There are those like Vandana Shiva whose reference points are, mostly, the lived reality of oppression. For others, such as Karen Warren, the starting point is the Western theoretical/philosophical tradition. Both are valid and necessary, and within each there are emancipatory strategies. In terms of language, a brief pause on some ecofeminist language may help illuminate the problem raised here between globalization and ecofeminism. Here is one example of an emancipatory strategy from Warren's edited book *Ecofeminism: Women, Culture, Nature*:

remything nature as a speaking bodied subject
erasing or blurring the boundaries between inner and outer landscapes, the self-other, human-non-human, I-Thou distinctions
re-eroticizing human relationships with a "bodied" landscape
historicizing and politicizing nature and the author as a participant in nature
expressing an ethic of caring friendship or a loving eye as a principle for relationships with nature
attempting to unseat vision or mind knowledge from a privileged position, positing the notion that bodies know.[34]

I am drawn into this beautiful, seductive, and soothing possibility. Yet, in front of the globalization ghoul, it feels powerless and irrelevant, even ludicrous. The ecofeminist affirmation of Life—and for religious ecofeminists, the sacred dimensions of Life—seems incommensurable in the face of globalization and its manipulation of Life: the hydra-headed forms of genetically altered food, plants, animals, and the Human Genome Project.[35] For example, over 80 percent of food in industrialized countries has been genetically modified. Of the countless patents on genes, there are at least fifty patents on the DNA of indigenous peoples. Life is a market commodity.

The above ecofeminist emancipatory vision and strategy seems weak in the context of today's world. It seems that ecofeminist visions, at times, speak of a reality that exists, but only as an imagined or dormant reality, a fleeting possibility. Rosemary Radford Ruether names this as a myopia in Northern ecofeminism, due to its emphasis on theory that doesn't make concrete connections with women at the bottom of the socioeconomic system. She writes, "we must recognize the ways in which the devastation of the earth is an integral part of an appropriation of the goods of the earth whereby a wealthy minority can enjoy strawberries in winter, while those who pick and pack the strawberries lack the money for bread and are dying from pesticide poisonings."[36] Northern ecofeminism can fall prey to cultural escapism, illusions, and irresponsibility. What if the problem is greater than myopia? What if the problem is located in a distortion in method and starting points, interlocutors, and a lack of attending to the world? It is disconcerting that the more time that is spent on developing forms of ecofeminist responses as above, the more powerless it will be in the face of globalization. Worse still, ecofeminism will be not only powerless, but participating in the destruction of the world while creating beautiful theories about alternative futures. Can the current forms of liberation theology methodologies, feminist theories, and ecofeminist efforts be effective in the face of globalization?

Outside of theological boundaries, there is clearly a rise in the numbers and effectiveness of citizen's groups, environmental and social networks, feminist coalitions, activists, and communities of resistance. There are other realities of resistance simmering on the margins. The growing public protests to globalization in Washington, Quebec City, and the World Social Forums indicate that some citizens are mobilizing. How can the resistance groups be heard more, and have greater voice and power? There are alternative teaching methods and signs of religious leadership outside academic clubs that are addressing globalization.[37]

Perhaps one of the problems is the limitation of religious academic discussions. For example, every year at the American Academy of Religion, where thousands of religious academics gather, less than ten of over two thousand papers given talk about economic issues. A similar ratio can be found in terms of courses and publications. Religion is not attending to globalization. Yet, to paraphrase Dan Maguire, if we are not addressing the economic, social, and ecological crises, religions are an obsolete distraction.[38]

Theology must get in the global game in a real and provocative way. It is my hope that the discourses of ecofeminist analysis and religious insights do not fall prey to fragmentation and that these discourses can move from isolated or abstract theoretical conversations into public arenas, into what Donna Haraway calls the "real-world" patterns of power and authority.[39]

Globalization is a force in history, determining and shaping cultural realities. There has never been such a hegemonic influence of this magnitude. If the alternative voices, such as that of ecofeminism, are to be meaningful, then they must contend with economic globalization. Only then will there be a meaningful exchange between ecofeminism and globalization.

Notes

1. Maude Barlow and Tony Clarke, *Global Showdown: How the New Activists Are Fighting Global Corporate Rule* (Toronto: Stoddart, 2001); Joshua Karliner, *The*

Corporate Planet: Ecology and Politics in the Age of Globalization (San Francisco: Sierra Club, 1997); David Korten, *When Corporations Rule the World* (West Hartford, Conn.: Kumarian Press, 1995).

2. Jordan quoted in Naomi Klein, *No Logo: Taking Aim at the Brand Bullies* (Toronto: Vintage, 2000), 267.

3. Marilyn Waring, *If Women Counted: A New Feminist Economics* (San Francisco: Harper Collins, 1988).

4. David Korten, *The Post Corporate World: Life after Capitalism* (West Hartford, Conn.: Kumarian Press, 1999).

5. Barlow and Clarke, *Global Showdown*, 75–76.

6. Beverly Wildung Harrison and Carol Robb, eds., *Making the Connections: Essays in Feminist Social Ethics* (Boston: Beacon, 1985), 245.

7. Harrison and Robb, *Making the Connections*, 245.

8. Larry Rasmussen, *Earth Community, Earth Ethics* (Maryknoll, N.Y.: Orbis, 1996).

9. Thomas Berry, *The Dream of the Earth* (San Francisco: Sierra Club Books, 1988); Thomas Berry, "The Seduction of Wonderworld," *Edges* 3 (1990): 8–14.

10. Rosemary Radford Ruether, *Women Healing Earth: Third World Women on Ecology, Feminism and Religion* (Maryknoll, N.Y.: Orbis, 1996).

11. Naomi Goldenberg, *The Changing of the Gods: Feminism and the End of Traditional Religions* (Boston: Beacon Press, 1979); Elizabeth Johnson, *She Who Is: The Mystery of God in Feminist Theological Discourse* (New York: Crossroads, 1993).

12. Catherine Keller, "Seeing and Sucking: On Relation and Essence in Feminist Theology," in *Horizons in Feminist Theology: Identity, Tradition and Norms*, ed. Rebecca Chopp and Sheila Greeve Davaney (Minneapolis: Fortress, 1997), 56–66.

13. Dorothee Sölle, *The Window of Vulnerability: A Political Spirituality* (Minneapolis: Fortress, 1990); Mary McClintock Fulkerson, *Changing the Subject: Women's Discourses and Feminist Theology* (Minneapolis: Fortress, 1994).

14. Barlow and Clarke, *Global Showdown*, 86.

15. Korten, *When Corporations Rule the World*, 111.

16. Vandana Shiva, *Reith Lecture 2000*, <http://news.bbc.co/uk/hi/english/static/vents/reith_2000lecture5.stm>; Vandana Shiva, *Biopiracy: The Plunder of Nature and Knowledge* (Toronto: Between the Lines, 1997).

17. Shiva, *Reith Lecture*.

18. Theo Colborn, Dianne Dumanoski, and John Peterson Myers, *Our Stolen Future* (New York: Plume, 1997), 107.

19. Linda McQuaig, *All You Can Eat: Greed, Lust and the New Capitalism* (Toronto: Penguin, 2001), 42–47.

20. Murray Dobbin, *The Myth of the Good Corporate Citizen: Democracy under the Rule of Big Business* (Toronto: Stoddart, 1998).

21. Bruce Harrison, *Going Green: How to Communicate Your Company's Environmental Commitment* (Chicago: Irwin Professional Publishers, 1993).

22. Andrew Rowell, *Green Backlash: Global Subversion of the Environmental Movement* (London: Routledge, 1996), 109.

23. Tony Clarke, *Silent Coup: Confronting the Big Business Takeover of Canada* (Toronto: James Lorimer and Co., 1997), 259.

24. Rebecca Chopp and Sheila Greeve Davaney, eds., *Horizons in Feminist Theology: Identity, Tradition and Norms* (Minneapolis: Fortress, 1997), 229.

25. Chopp and Davaney, *Horizons in Feminist Theology*, 224.

26. Mary Grey, communication with author, March 12, 2001.

27. Julia Kristeva, interview by Alice Jardine, in *Discources: Conversations in Post-modern Art and Culture*, ed. Russell Ferfuson et al. (Cambridge: MIT Press, 1990), 84, quoted in Catherine Keller, "Seeing and Sucking: On Relations and Essence in Feminist Theology," in *Horizons in Feminist Theology*, ed. Rebecca Chopp and Sheila Greeve Davaney (Minneapolis: Fortress Press, 1997), p. 64.

28. Beverly Harrison and Catherine Keller, "Feminist Liberation Theology for the Eco-Justice Crisis," panel at *Theology for Earth Community Conference* (New York: Union Theology Seminary, October 7, 1994).

29. Harrison and Keller, "Feminist Liberation Theology for the Eco-Justice Crisis."

30. Hewitt, *Critical Theory of Religion.*

31. Sölle, *Window of Vulnerability*, 153.

32. Berry, "The Seduction of Wonderworld," 8–14.

33. Geradl Karl Helleiner, "Market Politics and Globalization: Can the Economy be Civilized?" The 10th Annuel Raúl Prebish Lecture, presented at United Nations Conference on Trade and Development (December 2000), 4.

34. Karen Warren, ed., *Ecofeminism: Women, Culture, Nature* (Bloomington: Indiana University Press, 1997).

35. Donna Haraway, *Modest_Witness@Second_Millennium.FemaleMan©_Meets_OncoMouse™* (New York: Routledge, 1997), 246–47.

36. Ruether, *Women Healing Earth*, 5.

37. Barlow and Clarke, *Global Showdown*; Korten, *The Post Corporate World.*

38. Daniel Maguire, *The Moral Core of Judaism and Christianity: Reclaiming the Revolution* (Philadelphia: Fortress, 1993), 13.

39. Haraway, *Modest_Witness*, 230.

II

CHALLENGES TO ECOFEMINISM: CONCRETE CASES

T HE ESSAYS IN THIS SECTION provide glimpses of ecofeminist interactions with specific cultures and particular issues. Case studies from Kenya, Mexico, and India challenge certain basic ecofeminist assumptions. Each case study promotes a more nuanced ecofeminism and warns us against any simplistic understanding of a women-nature correlation.

3

Women and Sacred Groves in Coastal Kenya: A Contribution to the Ecofeminist Debate

Celia Nyamweru

Introduction

Anthropologist Celia Nyamweru's fieldwork in Kenya analyzes how men and women of the Mijikenda, a farming people, use the environment and express values about sacred groves. Celia shows how Islam, Christianity, and Western (secular) education erode indigenous belief systems. Her research also reveals that the claims made by some ecofeminists that women are more nature friendly than men do not apply here. Celia's essay cautions us against making general claims about women and their relationships to nature.

ALTHOUGH IT HAS BEEN STATED THAT "the formulation, control and dissemi-nation of ecofeminist beliefs are firmly under the control of white women,"[1] Third World women of color have figured prominently in the ecofeminist debate, sometimes as commentators, frequently through case studies of their activism. Notable in the first category is Vandana Shiva. Largely (although not entirely) through her writings, the activities of the "tree hugging" Chipko movement are widely known and frequently cited in the scholarly and popular literature. Shiva and others have used the Chipko movement and other, mostly Indian, case studies to support certain general-izations about the relationship of Third World rural women to the environ-ment. One of these is that women have their own particularly valuable body of traditional or indigenous environmental knowledge. Another is that women are spiritually and culturally linked to the natural world (forests, lakes, rivers, etc.) in ways that men are not. It is also suggested that women suffer more from environmental degradation than men do, are quicker to perceive its effects, and are the first to protest against it.[2] Together these assumptions create a picture of Third World rural women as being, by virtue of their gen-der, more sympathetic with the environment and more likely to manage its re-sources sustainably than are men.

In her account of the Chipko movement, Shiva also depicts rural Indian women as resisting the environmentally damaging initiatives of outside forces.[3] Examples of this are the World Bank social forestry projects. Accord-ing to Shiva, women have resisted such projects while their male relations are generally passive, if not active collaborators with the agents of change that bring about "maldevelopment."[4] Thus we can add to the above generalizations about the relationship between women and the environment the assumption that women resist globalization more than men do.

In the years since Shiva first wrote, her work has been subject to critical reeval-uation by scholars citing Indian and African examples, among them Rao,[5] Agar-wal,[6] and Jackson.[7] Agarwal observed that Shiva's "examples relate to rural women primarily from northwest India, but her generalizations conflate all Third World women into one category."[8] Jackson, who did fieldwork in Zim-babwe, points out that "women act as agents in both environmentally positive and negative ways"[9] and stresses the need "to unpack the idea that women's 're-sponsibilities' make them environment friendly. The responsibility to provide firewood for cooking a meal may lead a woman, when faced with a firewood shortage, to plant a tree but it may also lead her to pull up a wooden fence and burn it . . . or any number of alternative responses."[10] Recently Sturgeon has com-mented that "Third World women are used as natural resources for white ecofeminists without respecting the particularity of their lives and choices; they are reduced to a symbolic generality; and they are seen as 'closer to nature.'"[11]

This essay is based largely on data collected during two field seasons in Kilifi and Kawale Districts, Coast Province, Kenya, in 1996 and 1997. Four hundred sixty interviews were carried out, 294 with women and 166 with men. I use the results in an attempt, as Jackson said above, "to unpack the idea that women's 'responsibilities' make them environment friendly." I look at the uses made of the environment by men and women of the Mijikenda, a predominantly farming people, and the opinions they express about the value of the environment and the changes that have occurred during their lifetimes. In particular, I focus on attitudes to the sacred groves known by the Mijikenda ethnic groups as *kaya forests*. The word *kaya* is the Mijikenda word for a village or settlement. These forests originated several centuries ago, possibly as settlement sites of the Mijikenda, but today they represent almost the only areas of natural vegetation in a landscape that is being intensively developed for farming, settlement, mining, and tourism. They are sites of cultural and spiritual significance to the Mijikenda, even though the indigenous belief systems are being eroded by the influence of Islam, Christianity, and Western (secular) education. The *kaya* forests are cultural and physical spaces around which the interests of men and women, young and old, may coincide or diverge. They are contested landscapes over which local, national, and global interests compete for control of scarce natural resources.

In this essay I describe how Mijikenda women and men use the plant products of the *kaya* forests and the surrounding farmlands. In doing so I hope to test the assumption that women's ways are essentially more sustainable than those of men. I depict the Mijikenda women's spiritual and cultural links to the *kaya*. I explore whether Mijikenda women are quicker than men to recognize changes in the physical environment. Finally, I consider Mijikenda men's and women's roles in *kaya* forest conservation projects and show how the activities of a particular conservation organization have influenced village women's use of and attitudes toward the sacred groves. This should provide an illustration of how globalization, through the spread of internationally funded conservation initiatives, influences gender dynamics relating to control of environmental resources.

There seems to be considerable consensus about the sociocultural significance of the *kaya* forests. Parkin, in his account of one of the Mijikenda groups, the Giriama's beliefs and practices, describes their view of their *kaya* as "the source of their cultural essence and the moral safeguard against complete politico-economic encapsulation."[12] He goes on to describe the Giriama *kaya* as "sacred, yet it is also a fount of power which is used politically to defend the *Kaya* and the Giriama people but also to control them internally."[13] This power was and is largely exercised by men. As Willis says, "the presentation of the *kaya* should be seen rather in the context of the power of old men, with which it is clearly linked," and "[k]nowledge and care of the *kaya* is still in the hands of old men now."[14] In the Mijikenda's precolonial society the power to manage community affairs rested in councils of male elders. Among the Giriama the governing body met in the *moro*, a sacred area in the center of the *kaya*. The Giriama have other semi-secret societies, some of which are exclusively for men.[15] Women have their own society, the *kifudu* healing cult, which apparently has little to do with *kaya* matters.[16] Although women have little or no direct authority over *kaya* matters, they do have designated roles to play in some important *kaya* ceremonies. Such ceremonies respond to disaster (actual or foreseen), never to celebrations of happy events. As Parkin points out, "When people visit the *Kaya*, perhaps to take an oath or participate in a trial by ordeal, to acquire medical knowledge, to be blessed, or as one of a number summoned to cleanse it, it is always in a crisis occurring without warning." Parkin also emphasizes that the *kaya* is not a site of pilgrimage, unlike shrines and sacred places in many other cultures. The Giriama do not "in the least subscribe to a hope or myth of return. It is enough that the place itself, its forest, and its medicines and knowledge, remain in an uncontaminated state."[17]

During the second half of the twentieth century many of the *kaya* forests were drastically reduced as forest land was converted to cropland and buildings. A 1994 survey by the Coastal Forest Conservation Unit (CFCU) says that there are fifty-two *kaya* forests remaining in Kwale and Mombasa Districts and thirty-nine in Kilifi (and the new Malindi) Districts. Although the areas of many *kaya* forests have not been accurately surveyed, they range from less than twenty acres to several hundred acres, and the total area of *kaya* forest remaining is about four thousand acres. Some have been considerably degraded by human activity, while others remain largely undisturbed. Outsiders are not welcome in some of the *kaya* forests, while at others the elders are willing to take visitors inside. On a visit in August 1995 to three Kwale *kaya* forests we followed a narrow path through a thick outer wall of herbaceous plants, bushes, and creepers. Once through this, the forest floor was relatively open under the tall trees. Leaf litter and fallen branches contribute to a cool, moist, shady atmosphere. These oases of tranquility are surrounded by smallholder farms, mines, grazing land, settlements, commercial plantations and, in some places, the golf courses, bungalows, and swimming pools of luxury hotels catering to the international tourist trade that is an important source of foreign exchange for Kenya.[18]

During my field study in Kenya, I was able to interview women and men about their thoughts on the value of the *kaya* forest. Both put considerable emphasis on the value of the *kaya* forests as sources of timber for building and firewood; men stressed building poles and timber, and women stressed firewood. This is not surprising, given that Mijikenda women are responsible for providing the household fuel, while men build house frames and are carpenters and saw millers. The *kaya* forests are not the only sources of wood available to most families. From my interviews I learned that women also obtain firewood from trees on their own farms, from land belonging to other people, and from uncultivated (although probably individually owned) land. Women provided more details about the source of firewood than men did, as might be expected since they actually cut the wood and carry it home. Most women (68 percent) said they collected firewood from their own land; only 9 percent admitted to doing so from the *kaya* forest. In most of these villages women have access to a variety of trees and bushes, both wild and cultivated. Mango trees and cashew nut trees (both cultivated trees) are judged to yield good wood, while women also use waste material from coconut palms.

Given the very different labor inputs to firewood collection by women and men, it was somewhat surprising that their opinions on the difficulty of obtaining firewood were quite similar. Men of all age groups were as outspoken about the problems facing women due to firewood shortages as were women. Answers to this question did not differ according to gender but according to *CLASS?* village and family circumstances. One middle-aged male respondent reported that "for now I do not experience a major problem of firewood since my farm can supply enough firewood for me and my wife. Maybe for the years to come when I clear all my land for farming." A young woman from another village stated: "The other farm where we fetch our firewood is quite far, about an hour walk, thus the distance problem. Otherwise we must rely on the supply from coconut trees from around." When asked if she took firewood from the *kaya*, one young woman responded, "with me, no. But some who are near the *kaya* forest do fetch firewood."

The general question about the *kaya*'s value elicited little mention of nontimber forest products; however, when people were specifically asked about other *kaya* plant products, they provided a lot of information. I was particularly interested in women's knowledge and use of medicinal herbs. Herbal remedies are widely used by the Mijikenda, as by many other Kenyan ethnic groups. In most Mijikenda groups there are herbal remedies that are prepared and administered by healer specialists called *waganga*, while others are prepared and administered by the patient or by a family member. Both men and women may be *waganga*, and, within the family, women are responsible for diagnosing and treating many minor ailments. Women also play a significant role as diviners. A women's role as a healer may also result through her position in the *kifudu* healing cult mentioned earlier.

My initial assumption was that women would have greater familiarity with medicinal herbs than men and that they would gather most of these herbs from the *kaya* forests. I soon realized that most of the medicinal herbs named by the women came from wasteland and farmland around the houses or on the margins of cultivated fields. Eleven of the women interviewed at Kinondo stressed that useful plants were not found in the *kaya* forest. One woman at Tiwi distinguished between the herbal medicines known by the women, which are found around the homesteads, and the herbs in the *kaya*, known only by the *waganga*. Three women at Muhaka stressed that the *waganga* collected medicinal herbs from the *kaya*. Data from the interviews supported these preliminary results: men emphasized the *kaya* forest as the source of medicinal herbs, while women emphasized farmland. I asked whether women felt that it was important to teach children how to recognize and use medicinal plants, and their answers were unanimously positive. I also asked women about the need to protect medicinal herbs where they are growing, and some thought it was a good idea, but about half said that these plants spring up naturally after the rains so no protection is needed.

Another possible nonforest product is food, both fruit and wild greens. Wild greens are eaten by many Kenyans as a nutritious and tasty relish or sauce to accompany the staple starchy food. Women, who mentioned that they gathered many more specific greens than men, reported that they were collected from farmland. Men made more mention of wild greens from the *kaya* forest. Among those who specified a source for wild fruits, men emphasized the *kaya* forests, women emphasized farmland. Occasionally women mentioned other plant products from the *kaya* forests. Some said that sweet smelling flowers grow in the forest; one woman said she might pick them if she found them while collecting firewood. One or two of the women interviewed mentioned fiber plants and a wild root called *mariga*, which was edible after being soaked in water for two days.

Overall, these interviews suggest that the material value of the *kaya* forests to these people depends largely on wood products that are likely to be exploited on an unsustainable basis, both by women (firewood) and by men (building poles and timber). Nontimber forest products are not seen as important resources of the *kaya* forests by men or by women, and there is no evidence that women make more use of nontimber forest products than men do. Women emphasize farmland and the environs of the homestead as the source for wild greens, medicinal herbs, and even wild fruits. My interviews do not provide an image of the *kaya* forests as essential sources of food and medicines for village women. Nor, indeed, do they seem to be of great importance to village men in this regard, except possibly for some specialist healers. This is in striking contrast with the central Indian tribal groups for whom the forests are important sources of tubers, leaves, fruits, and roots used as food and medicine.[19]

The Mijikenda also value the *kaya* forests for shade, coolness and, above all, because trees are believed to attract rain. In response to a question on the goodness of the *kaya*, the answer "the trees attract rain" came up frequently; however, this answer came from 47 percent of the men and only 26 percent of the women. If the Mijikenda were to sing their version of the Chipko women's song, "What do the forests bear? Soil, water and pure air," the choir would likely be dominated by men's voices.[20] Shiva describes a technique of "lopping" oak trees that is passed from one generation of Himalayan women to the next: "Groups of women, young and old, go together to lop for fodder, and expertise develops by participation and through learning-by-doing."[21] The Mijikenda don't use the forests for fodder, but have they developed their own sustainable regimes? Hawthorne describes coppiced trees and bushes from the edges of *Kaya* Kambe, which could be a sign of some sustainable use, but Hawthorne thought this a sign of degeneration.[22] In fact, Hawthorne found that a lot of cutting was taking place and predicted total destruction of the forest canopy by the year 2000.[23] This has not been the case, but given the increasing demand for firewood from urban industrial operations and families, small-scale sustainable measures like coppicing are unlikely to be enforceable. An old woman from Ukunda, the most urbanized of the settlements in my study area, complained that "the firewood is becoming hard to get nowadays as people from the main road come to collect firewood here."

In considering the cultural, spiritual, and social reasons why men and women consider the *kaya* forests to have value, the most frequent answer related to the fact that the *kaya* forests are the sites of prayers and rituals that are believed to avert disasters, such as drought or epidemics. Both the men and the women who were interviewed recognized this. A middle-aged woman explained in some detail: "It is a place where people can go inside and pray their problems—if there is inadequate rain, they ask for rain to fall. If there is an epidemic, they pray so that it goes away. If there is a war coming, they pray and perform some rituals and give sacrifices to their Gods and the war does not break." Men were more likely than women to refer to the importance of the *kaya* forests as part of Mijikenda history and culture and to their importance as hiding places in times of war or danger. One old man stressed: "This is the origin of our culture . . . the *kaya* is the navel of the whole world."

Women were three times as likely as men to say that the *kaya* forest had no value, and over six times as likely to say that they did not know the value of the *kaya* forest. This was particularly true of young women with estimated ages of between eighteen and twenty-five years, among whom nearly 40 percent did not know the value of the *kaya* forest or said it had no use, while only 15 percent of young men, 16 percent of women over forty-five, and 2 percent of the oldest men answered in this way.

Although many of the Mijikenda women and men I interviewed show a generally positive attitude to the *kaya* forests, a number of responses demonstrated fear of some of the forests' attributes. According to an old woman, "if you cut wood in the *kaya* you will hear someone you can't see calling you. If you go on cutting wood, you will die." Another old woman said that "things from the *kaya* should not come into the village. If you build a house with timber from the *kaya*, it will fall down. Women did not and do not collect herbs from the *kaya*." A middle-aged woman, who was anxious that *kaya* ceremonies be held, stressed the importance of following correct procedures, such as entering the *kaya* forest without shoes: "You can die by being hit by something if you go in without knowing the customs." The young wife of the chairman of the elders of the *Kaya* Kinondo gave more details: "The *kaya* is a place to be afraid of. If you cut a tree with a machete in the *kaya* the machete will jump and cut your leg, and the wound will not heal until you give the elders ceremonial cloth to do a ceremony for you to heal. If women collect firewood from the *kaya* and cook with it, they can get very ill."

The *kaya* forests are seen as the homes of spirits. A middle-aged woman from Tiwi commented, "long ago there were devils in there." Spirits referred to as *vitsimbakazi* are believed to live in or around the *kaya* forest and to require small grass shelters to be built for them. An old woman described how at night during the season of the northern monsoon (the hot season, December to March) the *vitsimbakazi* come outside the house and say, "mother, I want to give birth, but I don't have a house." Other responses brought out the feeling that the *kaya* should be a dark place: a young woman complained that "trees have been cut, it is now open. It should be dark as it was before." A middle-aged woman described how *Kaya* Jibana "had many trees and darkness inside the *kaya*, but if trees are cut irresponsibly the place will not be good because there will not be enough darkness." Similarly, an old woman at Kinondo stressed the importance of conserving the *kaya* forest as it is supposed to be dark and if the trees were cut it would "shine brightly."

This contrasts markedly from Myers' discussion of civilizations that developed in the forested areas of Southeast Asia. Myers talks of "a sense of numinous awe associated with their forest homelands" and goes on to say: "In contrast to the folklore of temperate zones, which often regards forests as dark places of danger, traditional perceptions of forests in the humid tropics convey a sense of intimate harmony, with people and forests equal occupants of a communal habitat."[24] However, other scholars assert less benevolent attitudes. Gadgil and Vartak contrast the ferocious goddesses of sacred groves near Pune in India, who demand animal, even human, sacrifices, with the gods of the villages, "milder male gods like Maruti who live tamely in a temple and are happy with the offering of a coconut."[25] My sense from interviews with Mijikenda villagers is that many of those who have been converted to Christianity and Islam may have hostile attitudes to the *kaya* forests and to the elders, identifying them, as an eighteen-year-old male did, with "witchcraft." At the same time, men and women who retain elements of their indigenous beliefs do not see the *kaya* forests in a purely positive light. The forests are the homes of supernatural beings and the powerful magic of the charms that are buried there and controlled by the presiding elders. They are seen as places of potential danger, only to be entered on sanctioned errands and according to prescribed behaviors.

If the *kaya* forests are not seen as intrinsically benevolent, are they seen as intrinsically female? Ortner authored an early critique of the assumed link between women and nature. She answered the question "is female to male as nature is to culture?" by observing that "woman is not 'in reality' any closer to (or further from) nature than man . . . but there are certainly reasons why she appears that way."[26] Even today, some ecofeminists see nature as specifically female and posit an intrinsic link between women and trees or forests. Yet, when we look at the practices and beliefs of different cultures, we can distinguish many versions of the gender/nature relationship. As Jackson states, "although for some societies there is evidence that women are associated with nature, men are too, and men are not exclusively associated with culture."[27] I did not specifically ask whether the *kaya* forests were women's places, but all the evidence about rituals and decision making related to *kaya* matters suggests that this is not the case. One middle-aged woman I interviewed at Kinondo felt that it was good that people should be kept out of the *kaya* forest, in particular women and young children. She explained that menstruating women pollute the *kaya* and, if she were to go and collect firewood from the *kaya* forest while menstruating, she would come home and find her children sick. The limitations, noted in the 1980s, on women entering *Kaya* Fungo, one of the *kayas* where the traditional beliefs are strongly retained, appear to have been even more rigid. In May 1988, a diviner, the son of a *kaya* elder, was told by one of his possessor spirits that *Kaya* Fungo was contaminated by evil for the following reasons:

> People had been entering the sacred place wearing shoes; Europeans (tourists) had been coming to the *Kaya*; goats and cattle had been slaughtered there (with normally only the pure and peaceful ram allowed to be sacrificed in cleansing and fertility ceremonies); radios had been played in it; people had worn shirts while visiting (only seamless clothes should be worn, such as a traditional cotton wrap); and that women other than the wives of *Kaya* elders (who are permitted) had also entered.[28]

It is easier for women (and Europeans) to enter other *kayas*; however, in all cases, Mijikenda women's relationships with the *kaya* forests appear to be controlled by men. Decisions on when to carry out *kaya* rituals are made by the elders. I was told by several woman at Kinondo that they were anxious for a *kaya* ceremony to be held, but "Mzee Abdallah does not like women's orders." A number of women were able to give me considerable amounts of detail about the rituals, and, from their narratives, it appears that these women had significant responsibilities. Women cleared the paths into the *kaya* forests, cooked the ceremonial meals, and took part in the dancing that went on throughout the night. The women did not enter the most sacred areas or take part in the prayers; this was restricted to the male elders. The home-based, female-controlled *kifudu* practices can be contrasted with the rituals of the *kaya*, which are set apart from the homestead, controlled by men, and deal with broader issues of communitywide well-being.

If women are, as some ecofeminists maintain, closer to the natural world, both spiritually and materially, then one might expect them to be quicker than men to recognize changes in the physical environment. To test this, I asked several questions that invited people to consider aspects of environmental change, from particular issues (water and firewood), to farming, to an open-ended question: "What environmental changes have you seen in your lifetime?" Most people stated that environmental change had occurred during their lifetimes, although men were slightly more likely to do so than women. The changes they described were overwhelmingly negative, except in a few cases. The people from Kauma and Ukunda villages mentioned the improved water supply in their communities, and a handful of people described positive socioeconomic changes, such as the old man who told me that "people did not use to construct permanent buildings but now people have money to build better houses." The most common responses from both men and women described lower or less reliable rainfall. This was followed by "crop yields have decreased," and the related response, "there is famine, there is less food." Since over 95 percent of both men and women said they were farmers, this concern is not surprising. In 1996, 21 percent of the women interviewed mentioned poor rains or drought as a problem facing them in farming, and the following year 24 percent of both men and women did so. Men and women were equal in agreeing that changes in farming had occurred since they were young, and that most of these changes were negative. The possibility that the rains will fail, resulting in crop failures and starvation and the need, therefore, to be able to do rituals in the *kayas* that avert this danger, are important elements of Mijikenda culture.

My sense from these results is that there is no major gender differential in the recognition of environmental problems and changes. Women are not shown to be more aware of environmental changes or to have more complaints about most environmental problems. In fact, in most cases, men tended to have rather more to say on these topics than women did, although, in part, this may have been because men felt more at ease in an interview situation and thus were more willing to air their complaints about the world.

I turn now to exploring men's and women's roles in *kaya* conservation. To what extent have women in the communities around the *kaya* forests been involved in activities against environmental degradation and for the preservation of the forests? To answer this question, it is important to examine the history of the *kaya* conservation movement, whose origins can be traced back to the 1980s. It was then that Western botanists began to take sustained interest in the *kaya* forests because of their biological diversity, and issues related to their conservation were raised in the global scientific community. Most people who visited the *kaya* forests in the 1980s observed that the influence of many of the male *kaya* elders' groups had significantly diminished. This had led to increased destruction of the forest by local community members extending their fields or cutting building poles and planks,[29] as well as by outside interests, such as the mining operations at *Kaya* Kambe and *Kaya* Mrima.[30] Hawthorne and his fellow students from Oxford University recommended that conservation of the *kaya* forests be achieved "through the traditional hierarchy of authority, either by assigning some of the elders local positions of authority, or by extending to them certain powers, perhaps equivalent to those of forestry officers or game wardens."[31] Robertson's accounts of her visits between 1982 and 1987 include descriptions of meetings and discussions with *kaya* elders. One example is her account of a visit to *Kaya* Ribe in Kilifi District:

> We held discussions with the wazee [elders] on 11 February 1987 and found that they were very concerned about the safety of the forest and asked for help in looking after it. They said there were two main threats. One was tree cutting by some people who sell planks to contractors in Mombasa. This had only begun recently within the last year but was difficult to stop without employing *askaris* [security guards] who could catch the offenders and take them to the Chief to be prosecuted.[32]

Robertson's discussions about the status of the different *kaya* forests were with male *kaya* elders, and she has little to say about women's conservation initiatives. The only mention of women that I can locate in the early literature is her description of a visit to Chonyi in May 1987 to talk to a local primary school and a women's group about a tree-planting project on the edge of *Kaya* Chonyi. The secretary to the women's group, a Mr. Stephen Mwandongo, showed her the tree seedlings that had been delivered, and she mentions that there was some local resistance to this project.[33]

Since a general consensus formed that the conservation of the *kaya* forests depended on working with the male *kaya* elders, this became an important element of the activities of the Coastal Forest Conservation Unit, which began its work in the early 1990s. CFCU is a unit of the National Museums of Kenya, part of the Ministry of Home Affairs and National Heritage (this has varied to some extent with reorganization of Kenya government ministries). CFCU has considerable independence since it has its own funding from the British Overseas Development Authority and the World Wide Fund for Nature, UK. At times it functions like a non-government organization, challenging local and central government officials and policies.

CFCU's project proposal committed the organization to "regular dialogue, planning and monitoring with the village *Kaya* committees and other village groups."[34] It also works with schools and youth groups to educate young people about the cultural and ecological value of the *kaya* forests through essay competitions, visits to schools, and field trips. The work with schools brings the *kaya* conservation message to girls as well as boys (almost all Kenyan elementary schools are mixed sex, although in most regions boys tend to outnumber girls, particularly in the higher grades). However, CFCU's project proposal did not include any specific goals of working with adult women, and, as it turns out, its activities in the villages around the *kaya* forests have often tended to enhance the power of the male elders in ways that do not always coincide with women's interests. CFCU encourages the elders to take action against violators of *kaya* regulations, either by fining them according to traditional sanctions, or by pushing for formal prosecution in local magistrates' courts. Some of those who are on the receiving end of these sanctions are women or girls who are caught collecting firewood in the *kaya* forests. As reported by CFCU in late 1997: "A group of young girls were intercepted cutting firewood in the *Kaya* [Kambe] and required to pay a fine of seven goats although this was later reduced to a token amount of Ksh 700, which, by local standards, is still a sizeable amount. The fine was paid and the girls' cutting tools returned."[35]

There is no doubt that the firewood and building poles issues have created considerable ambivalence about the conservation of the *kaya* forests. Of the people interviewed in 1997, over 40 percent of both men and women stated that there were problems in obtaining firewood. Respondents were evenly divided between those who attributed this to the activities of the *kaya* forest guards enforcing conservation regulations, and those who pointed to other causes, such as population growth or an increase in the area of cultivated land. Of those who believed that the difficulty in getting firewood was due to *kaya* conservation activities, some were aware that these regulations were not a recent development. A sixty-one-year-old man and an eighteen-year-old man said that "it has been prohibited since time immemorial." Another middleaged man put it this way: "The *kaya* is prohibited by our forefathers. I am neighboring the *kaya*, and I have never even attempted to go into the *kaya.*" A few respondents had a more complex story to tell—a very old woman said, "the *kaya* is prohibited but in those days we had some limits on the *kaya* where to fetch firewood. We had some zones in the *kaya* to fetch firewood." Those sentiments were echoed by a fifty-nine-year-old man: "During those days the elders would allow villagers to collect firewood at a low scale." These responses show that some people are aware that access to firewood was limited under indigenous control of the *kaya* forests. A few people identified outside agents: A middle-aged woman reported that "a European bought this *kaya*, and now we are not permitted to get firewood in there any more." A seventy-year-old man (and a member of a *kaya* committee) said that "these problems came in when CFCU started. Otherwise we used to obtain firewood from the *kaya.*" However, the majority of those interviewed saw it in terms of a simple contrast between past and present. "The changes are that we used to rely on the *kaya* but it is now prohibited," said a thirty-year-old woman.

Interviews demonstrated that women in several villages feel strong resentment about the denial of access to firewood, whether they blame it on the *kaya* guards (men from their own communities), on the *kaya* elders (also from their own communities), or on outsiders (Europeans or the CFCU). At the same time that many women, especially older women, recognize the cultural and spiritual value of the *kaya*, many women would like to see increased access to the forests as a way to ease the firewood problem. Although these attitudes are not exclusive to women, they are clearly strongly felt by women, and these attitudes do not lend themselves to the ecofeminist perspective that places all rural women, everywhere in the Third World, in the forefront of environmental conservation.

Over the years the CFCU personnel have forged an effective alliance with many of the *kaya* elder groups. The *kaya* elders have been given access to a wide forum, including regional meetings and, for a few, trips to biodiversity conferences and workshops in Nairobi. The committees of elders also have access to patronage; they appoint the *kaya* guards and may receive monthly salaries from the CFCU. Some have become quite sophisticated in their manipulation of the local media and in their interaction with the increasing numbers of visiting scholars, conservationists, and civil servants who visit the villages and, at times, the *kaya* forests under the auspices of CFCU. None of this has translated into power or influence for women, who remain on the margins, complaining about the prohibitions on collecting firewood that are enforced, sometimes inefficiently and sporadically, by the *kaya* guards who are answerable to the male elders. The influence of the global conservation movement, through the agency of CFCU, has provided the male elders committees with access to funds, advice, and encouragement, but it has not worked to bring village women into the conservation arena.

During a recent visit to coastal Kenya in August 2000 I learned of an initiative that suggests that this may be changing. In an attempt to widen community support for *kaya* conservation, CFCU is encouraging the establishment of *Kaya* Forest Community Conservation Groups. Membership will not be restricted to *kaya* elders and will include women and younger people. As of August 2000 four such groups had been constituted and had held formal meetings. The minutes of the first meeting of the *Kaya* Kauma conservation group list twenty-seven people present, of whom four were women; the *Kaya* Chivara group had thirty-three members, of whom seven were women. The creation of these groups is an important step toward involving more community members in *kaya* conservation, and, specifically, more women. However, given the deeply patriarchal and gerontocratic nature of Mijikenda society, the small numbers of women involved in these groups so far, and the tendency of women to say little in a formal setting where men are present, it may be a while before women are able to voice their concerns about access to *kaya* forest resources, and be heard.

The results of my interviews show that the relationship of Mijikenda women to the *kaya* forests is very different from the kind of ecofeminist image portrayed by Shiva and Mies. Among the Mijikenda, women do not make more use of nontimber forest products than men do, and both women and men extract forest products in a nonsustainable way. Women collect firewood from the *kaya* forests and men cut building poles, even though they recognize the sacred and cultural significance of these forests. This puts Mijikenda women in a very different relationship to nature from that of the Himalayan women described by Shiva, who were actively engaged in attempts to conserve forests that they did *not* recognize as sacred.

The perception of the *kaya* forests in Mijikenda culture cannot be interpreted in terms of benevolent space or of specifically female space. They are places of potential danger that have to be approached according to proscribed behaviors and to which women's access is strictly controlled. The rituals and prayers, which are a key part of the *kaya* forests' functions, are largely controlled by men. And even though economic and cultural forces, both global and local, have increased the risks to the *kaya* forests, women showed no greater recognition than men of those risks. Conservation initiatives since the early 1990s have focused mainly on the male elders, and in this context the impact of global conservation agencies has enhanced the role of men in *kaya* conservation, possibly at the expense of women. A new initiative by the CFCU may serve to balance this and to bring women into the ongoing struggle to conserve these unique cultural and biological resources.

Acknowledgments

I am extremely grateful to the personnel of CFCU for hospitality, transportation, and constant encouragement in my work; to St. Lawrence University for providing financial support for some of my fieldwork; and to those who read and commented on drafts of this chapter. Above all, I am grateful to the women and men of Kwale and Kilifi districts for sharing so much of their lives and their opinions with me—*asanteni sana*.

Notes

1. D. E. Taylor, "Women of Color, Environmental Justice, and EcoFeminism," in *Ecofeminism: Women, Culture, Nature*, ed. Karen J. Warren (Bloomington: Indiana University Press, 1997), 70.

2. Maria Mies and Vandana Shiva, *Ecofeminism* (London: Zed, 1993), 2–3.

3. Mies and Shiva, *Ecofeminism*, 73.

4. Mies and Shiva, *Ecofeminism*, 79.

5. B. Rao, "Dominant Constructions of Women and Nature in Social Science Literature," *Capitalism, Nature, Socialism*, Pamphlet 2 (New York: Guilford Publications, 1991).

6. Bina Agarwal, "The Gender and Environment Debate: Lessons from India," *Feminist Studies* 18, 1 (spring 1992): 119–58.

7. C. Jackson, "Women/Nature or Gender/History? A Critique of Ecofeminist 'Development,'" *Journal of Peasant Studies* 20, 3 (April 1993): 389–419.

8. Agarwal, "Gender and Environment Debate," 124–25.

9. Jackson, "Critique of Ecofeminist 'Development,'" 413.

10. Jackson, "Critique of Ecofeminist 'Development,'" 412.

11. Noël Sturgeon, *Ecofeminist Natures* (New York: Routledge, 1997), 128.

12. David Parkin, *Sacred Void* (Cambridge: Cambridge University Press, 1991), xv.

13. Parkin, *Sacred Void*, 12.

14. Justin Willis, *Mombasa, the Swahili, and the Making of the Mijikenda* (Oxford: Clarendon Press, 1993), 38–40.

15. A. M. Champion, "The Agiryama of Kenya," ed. John Middleton (London: Royal Anthropological Institute, 1967).

16. M. Udvardy, "Kifudu: A Female Fertility Cult among the Giriama," in *The Creative Communion: African Folk Models of Fertility and the Regeneration of Life*, ed. A. J. Widding and W. van Beek (Stockholm, Sweden: Almqkvist & Wiksell International, 1990): 137–52.

17. Parkin, *Sacred Void*, 10.

18. Celia Nyamweru, "Sacred Groves Threatened by Development," *Cultural Survival Quarterly* (fall 1996): 19–21.

19. Vandana Shiva, *Staying Alive: Women, Ecology and Development* (London: Zed, 1989), 59–60.

20. Shiva, *Staying Alive*, 77.

21. Shiva, *Staying Alive*, 66.

22. W. Hawthorne et al., "Kaya: An Ethnobotanical Perspective" (University of Oxford: Department of Botany, 1981), 60–68.

23. Hawthorne et al., "Ethnobotanical Perspective," 66.

24. Norman Myers, *The Primary Source: Tropical Forests and Our Future* (New York: W. W. Norton, 1984), 13.

25. Madhav Gadgil and V. D. Vartak, "The Sacred Groves of Western Ghats in India," *Economic Botany* 30 (1976): 156.

26. Sherry B. Ortner, "Is Female to Male as Nature Is to Culture?" in *Women, Culture, and Society*, ed. M. Z. Rosaldo and L. Lamphere (Palo Alto, Calif.: Stanford University Press, 1974), 87.

27. C. Jackson, "Critique," 396.

28. Parkin, *Sacred Void*, 29.

29. Hawthorne, "Ethnobotanical Perspective," 60.

30. S. A. Robertson, "Preliminary Floristic Survey of Kaya Forest of Coastal Kenya," *Report to the Director* (National Museums of Kenya, 1987), 43, 96.

31. Hawthorne, "Ethnobotanical Perspective," 95.

32. Robertson, "Floristic Survey," 49.

33. Robertson, "Floristic Survey," 49.

34. Coastal Forest Conservation Unit, Funding Proposal to ODA-JFS (Nairobi, 1994), 5.

35. Coastal Forest Conservation Unit, *Quarterly Report XIV for the period 1 October–31 December 1997* (Kilifi, 1998), 8.

4

Indigenous Feet: Ecofeminism, Globalization, and the Case of Chiapas

Lois Ann Lorentzen

Introduction

Lois Ann Lorentzen, social ethicist, examines the complex interactions among women, nature, and culture in Chiapas, Mexico. Here the women-nature construct functions differently than it is understood within most ecofeminist theory. Lois shows that the cultural patterns and specific sociopolitical challenges faced by women in highland Chiapas cannot be understood within a narrow ecofeminist framework. These research findings reveal not only what is occurring in Chiapas, but also that there is no one understanding of the interrelations between women and nature. Each context requires in-depth consideration.

Indigenous feet tread the soft earth in awe and respect.

Ariel Salleh, *Ecofeminism as Politics*

ECOFEMINISM CLAIMS TO BE the "third and international wave of feminism."[1] As environmental devastation continues at a ferocious pace, as global capital stretches its reach around the planet, and as the world's poor continue to suffer, a theoretical perspective that links forms of oppression is difficult to resist. As both a feminist and an environmentalist, the theory and practice of ecofeminism held special appeal for me. I embraced the term and its principles and then "discovered" its existence in others, including indigenous women. In several articles I have claimed that given women's roles in providing family sustenance in many less affluent nations, women possessed a space that was neither nature nor culture, and, in the process, I questioned such a dichotomy.[2] In other articles, I explored ecofeminism in Central American, especially Salvadoran contexts.[3] I still believe, as I have argued elsewhere, that it is empirically demonstrable that women generally suffer the impact of environmental deterioration more than men, and in some cases have higher levels of environmental activism.[4] Following sabbatical time spent in Chiapas, Mexico, however, I have come to question concepts I once held dear. This essay explores certain basic principles found in ecofeminism; the discourse concerning indigenous and Third World women used by ecofeminists from more affluent nations; crosscultural perspectives on ecofeminism; and the usefulness of ecofeminist principles and practice for political and social movements in the face of globalization, when applied to a specific context, that of Chiapas, Mexico.

As Karen Warren notes, ecofeminism is an "umbrella term" for a wide variety of perspectives, some of which are compatible with each other and others that are contradictory.[5] What holds these disparate positions together is the claim that "there are important connections between the unjustified dominations of women, people of color, children and the poor, and the unjustified domination of nature . . . ecofeminism is about interconnections among all systems of unjustified human domination—beginning with gender as a category of analysis."[6] Ecofeminism provides both a "theory of domination and strategy for change" and thus is both a social movement and a theoretical perspective.[7] Ariel Salleh answers the question, "Who is an ecofeminist?" by stating that an ecofeminist is "a man or woman whose political actions support the premise that the domination of nature and domination of woman are interconnected."[8] Noël Sturgeon defines ecofeminism as a "movement that makes connections between environmentalisms and feminisms, more precisely it articulates the theory that the ideologies that authorize injustices based on gender, race, and class, are related to the ideologies that sanction the exploitation and degradation of the environment."[9]

IS PATRIARCHAL CONTROL LINKED TO DEGRADATION OF NATURE??

I am not going to review the wide variety of ecofeminist theories and practices covered by this umbrella term. Ecofeminism concerns itself with multiple issues, attempts to be globally oriented, and draws from a wide array of intellectual traditions. As Warren notes, ecofeminists may explore the woman/other human others/nature interconnections by analyzing historical (typically causal), conceptual, empirical, socioeconomic, linguistic, symbolic and literary, spiritual and religious, epistemological, political, and ethical interconnections.[10] Ecofeminists may be Marxists, socialists, cultural ecofeminists, postcolonialists, postmodernists, goddess-worshipers, and more. I applaud this robust theoretical pluralism. What concerns me is what ecofeminists claim to hold in common. How helpful are ecofeminism's basic *QUESTION* principles when applied to a specific community and context, that of indigenous women in the highlands of Chiapas, Mexico?

Karen Warren writes, "What *all* ecofeminists agree about, then, is the way in which the *logic of domination* has functioned historically within patriarchy to sustain and justify the twin dominations of women and nature."[11] While recognizing that there are

> a variety of different feminist perspectives on the nature of the connections between the domination of women (and other oppressed humans) and the domination of nature, "Ecological feminist philosophy" is the name of a diversity of philosophical approaches to the variety of different connections between feminism and the environment.[12]

And also, "What all ecofeminist philosophers *do* hold in common . . . is the view that there are important connections between the domination of women (and other human subordinates) and the domination of nature."[13]

What I want to do, using the case of Chiapas, is question this central and core premise of both philosophical and religious ecofeminism. I will do this *✳ PURPOSE* by challenging the notion that woman is conceptually linked with nature and man with culture, and by questioning the notion that patriarchal control of women is necessarily linked to the oppression of nature.

A central claim for many is that the conceptual basis of the linked dominations of women and nature stems from hierarchically organized value dualisms such as culture/nature, human/nature, man/woman, which associate nature and woman as inferior to culture, to the human, to the male. In the case of Chiapas, it is simply not true that women are overidentified with nature. For the indigenous people of San Pedro de Chenalhó for example, both men and women are people of the corn. For nearly all indigenous Chiapans, care of the *milpa*, the land needed to grow corn, beans, and squash, is viewed as male work. As Christine Eber writes of Pedranos, "True men care for the land, plant and harvest corn. True women turn corn into food for the family."[14] For the Pedranos, the male force is found in the earth, the force that cares for the *milpas*. In another village, San Pedro de Chamula, men care for the land and the crops, whereas women care for the sheep. Thus it is that those who care most directly and intimately for the land itself are the men.

Conceptually, the metaphor of sun/moon translates to male/female; heat/cold; courage/timidity; strength/weakness. As Brenda Rosenbaum notes in her excellent ethnography, *With Our Heads Bowed: The Dynamics of Gender in a Maya Community*, this construction provides an "essential framework for judging behavior as masculine and feminine."[15] The point is that the pairing male/female does not correspond to culture/nature. This does not mean that women are not oppressed; it means that the culture/nature dichotomy does virtually nothing to illuminate this particular form of gendered oppression. If anything, men are *more* associated with nature than are women. Symbolic justification for female subordination arises from the *type* of nature with which men and women are associated. For example, for Chamulans, men are associated with corn whereas women are associated with potatoes. Corn grows tall and is considered superior to potatoes. Just as the moon follows the sun, so must women walk behind men on the trail. The sun is hot whereas the moon is cold and therefore women are more vulnerable and less courageous than are men.[16]

Not only are men linked conceptually with nature in the Chiapan case, women are more likely to be identified with culture. Indigenous women are seen as bearers of traditional culture given their roles in weaving, pottery making, and fruit collecting. Women save the language, the rituals, and traditions. Men, more likely to leave their villages to seek paid employment, often symbolize compromise with outsiders, especially with the hated Ladino.[17] Positively, this role as culture's preservers may give women increased local status. Yet women's perceived role as guardians of culture may place them at risk in actual conflicts as the government pays increased attention to women in its fight for symbolic control of traditions and customs. The Mexican government is in a paradoxical position of both valuing and being threatened by indigenous women and their role as culture's preservers. They must value them because use of indigenous traditions is part of the state's construction of national identity. The rescue of cultures and customs is seen as a source of resources—think of the numerous *ballet folclórico* groups and the use of Mayan and Aztec symbols in official depictions. The exoticizing of the indigenous, and especially the indigenous woman, is certainly necessary for tourism. Shockingly, tourism in Chiapas increased dramatically following the Zapatista uprising, in part because of the government's *Mundo Maya* (Mayan World) tourism campaign in which the indigenous (especially the indigenous woman) stands for Mexican culture. The timing of the *Mundo Maya* tourist campaign and Zapatista success should not surprise us. Chiapas faced the twin pacification strategies of tourism and militarization.

Yet, official indigenous discourse masks the threat the government perceives in these culture-bearing women. In the ongoing state of low-intensity military conflict that characterized day-to-day life in Chiapas through much of the 1990s, the bodies of women literally became material for an official nationalism while at the same time women were considered dangerous cultural transmitters. Rape and threats of rape became all too common weapons used by the counterinsurgency in the Chiapan zones of conflict.

Men identified equally or more with nature than women, women identified by both the state and the community as bearers of culture, subordination of women by men justified symbolically by identifications with parts of nature rather than with culture, patriarchal control of women that does not translate into exploitation of nature, patriarchal control of women that seemingly is not linked to a larger logic of domination—how does theoretical ecofeminism help illuminate this reality? It may be the case, as Janet Biehl suggests, that "systems of domination have their own logic."[18] Irene Silverblatt carefully demonstrates, in *Moon, Sun, and Witches: Gender Ideologies and Class in Inca and Colonial Peru*, how the Inca, like the Maya, both past and present, had social systems characterized by hierarchical domination but not of domination of nature.[19] If I cling to my notions of ecofeminism, I cling to an explanatory mechanism that hinders more than helps us understand Chiapas. It is striking that in the Women's Revolutionary Laws formed by Zapatista women and affirmed at the Indigenous Women's Congress in Chiapas in 1997, no mention was made of nature, of women's connection to nature, or of the impact of environmental deterioration specifically on women.[20] Why didn't they mention nature? False consciousness? Hardly likely given their otherwise sophisticated set of demands. Or perhaps these feminist indigenous women did not mention nature because the woman/nature/oppression connection is just not there for them. Bina Agarwal's concerns are relevant here. Agarwal has written that the Chipko movement reflected women's participation in a peasant movement rather than in an explicitly feminist movement.[21] It may be that the women of the Women's Indigenous Congress perceive that their environmental concerns relate to their peasant and Zapatista struggle, not to their feminist activism. Both are compelling, but they may not be perceived as connected.

To be fair, many ecofeminists do note that the interconnections between women and nature may be specific to Western thought and culture. Karen Warren's last book is quite nuanced in this respect. In her preface she writes that the "central conceptual issue concerns the nature of the interconnections, at least in western societies, between the unjustified domination of non

[handwritten margin notes:] QUESTION

* DON'T SEE ENV. STRUGGLE AS RELATED TO GENDER BUT RATHER CLASS

human nature and other human others on the one hand and the unjustified domination of non human nature on the other."[22] However, although this may be a Western construction, it still has global effects. This is because, as Plumwood writes, the "construction of certain categories of humans as nature has naturalized their domination."[23] Chiapan women may not be more like nature within their own group, but as indigenous, the dominant culture both naturalizes and feminizes them. Warren writes that ecofeminists

> begin with gender as a category of analysis not because gender oppression is more important than other forms of oppression. . . . It is because a focus on "women" reveals important features of interconnected systems of human domination: First, among white people, people of color, poor people, children, the elderly, colonized peoples, so-called Third World people, and other human groups harmed by environmental destruction, it is often women who suffer disproportionately higher risk and harms than men.[24]

As colonized peoples, indigenous women are naturalized, feminized, and therefore dominated. As women they suffer disproportionately from environmental harm. It is patriarchy, in this case, capitalist patriarchy that is the macrohistorical source of the twin dominations of women and nature that these Chiapan women experience, according to these ecofeminist claims.

Yet even the preceding claims are problematic. What about globalization as the most recent manifestation of capitalist patriarchy? There are three primary ways in which globalization and the global economy affect the lives of Chiapan indigenous peoples: tourism, increased telecommunication, and the impingement of world markets as represented by NAFTA. All are more complicated than they seem at first. Globalization is evidenced by the increased tourism I mentioned earlier. However, in an odd twist, increased tourism has allowed women to continue in their roles of culture bearers. Women, through textile and artisan cooperatives, through selling in the markets of San Cristóbal de las Casas, have been able to bring in needed income to households while at the same time practicing their traditional skills. Rosenbaum claims that a certain level of tourism in Chiapas actually serves to preserve indigenous culture. Women providing income through weaving or pottery making is seen as preferable to serving as a maid in a Ladino home or leaving home to work in a factory. Globalization as represented through information technologies has actually aided the Zapatista cause, allowing for a level of international support, solidarity, and visibility unprecedented in Chiapan history. The current Zapatista uprising is only one in a history of indigenous resistance movements in southern Mexico, but it is certainly the most visible. Here, indigenous peoples (or more accurately, some leaders) use e-mail to protect traditional ways of life.

Finally, NAFTA, as the symbol for one version of dominant global systems of economic profit, certainly seems relevant in the Chiapan context. It is clear that NAFTA and the subsequent erosion of communal ownership patterns, as well as the encouraging of export crops, are destructive in Chiapas. However, the primary oppression has been due to internal rather than external logics. By 1847 the Mexican government decreed that indigenous peoples had to live in pueblos, forcing them to work for *patrones*. This initiated an essentially feudalistic economic system in Chiapas, especially in the highlands, which remains so up to the present. Internal markets and the wealth of the hacienda owner formed the primary logic. By 1911 Ladinos had taken most of the land in Chiapas. Thus, in highland Chiapan stories and rituals, a common theme exists—Ladino domination of indigenous peoples. Chamulan ideology depicts the Earth Lord as a supernatural Ladino. The Earth Lord's wife seeks servants, and both the Earth Lord and his wife try to capture the souls of Chamulans. Chamulans who envy their neighbors are said to have struck a deal with the Earth Lord. In the realm of the Earth Lord, Chamulans work as peons and servants. Those who adopt Ladino values risk eternal damnation.[25] These powerful rituals directly identify Ladinos, who may or may not be part of the circuits of the global economy, as the cause of indigenous pain, suffering, and poverty.

During fiestas and festivals in the highlands, Ladinos are ridiculed in rituals. According to Eber, the men who receive the most laughter during carnival are those who impersonate either Ladinos in ridiculous situations or women, especially sexually transgressive women. Women's transgressions are related to their weakness and sexual promiscuity, whereas Ladinos are ridiculed for their undeserved domination.[26] For women, the rituals serve to discourage deviations from traditional gender norms. After all, "this is how the ancestors did it."[27] If ecofeminists analyze linguistic, symbolic, and religious interconnections between women and nature to understand oppression, these rituals are problematic. The Ladino husband and wife are equated with the Earth as Earth Lord. This Earth being dominates the indigenous generally, yet the Ladino is also ritually ridiculed for this oppressive behavior. Indigenous women, who are not linked more with nature, also face ritual ridicule due to the "fact" of their weakness and sexual promiscuity. Control and domination are rejected and interrogated, as well as affirmed in the juxtaposition of these rituals.

I am in no way saying that the imperatives of the global economy do not make life worse for many Chiapans. It is clear from Zapatista analysis that the global market has a destructive impact on indigenous peoples. The production of maize, sacred to many indigenous groups, has decreased as the United States floods Mexico with cheap corn. Coffee production has been seriously

affected and more and more indigenous from throughout southern Mexico
have been forced to leave their lands, flooding already overcrowded border
towns. However, to leave it at that level of abstraction does little to illuminate
the ongoing Ladino/indigenous struggle and the particular forms it takes.
Here we have intertwined but separate histories, multiple levels of oppression
that may or may not be connected. There are oppressions internal to indigenous communities, the Ladino/indigenous struggle (which, for most Chiapans, is primary), the state/indigenous struggle, and the increasing demands
of the global economy. In terms of day-to-day life, however, I suspect that
most indigenous would still rather work in a *maquila* (foreign-owned factory)
than in the hacienda of a Ladino *patrón*.

[handwritten margin note: BUT THIS DOESN'T MEAN THAT FOREIGN OWNERSHIP IS BETTER]

The Zapatista uprising provided opportunities for indigenous women to
demand changes in gender roles within indigenous communities. According
to Neil Harvey, these changes were most apparent in the Lacandon forest
where "three interrelated processes helped indigenous women to assert their
own demands during 1994–96." The processes, according to Harvey, were:

1. The very fact of colonization itself, which required women to adopt
   nontraditional roles in the new lowland *ejidos*. Due to the lack of government assistance . . . the migrants were left to clear forest on their own.
   Women carried out as much of this work as the men.
2. The incorporation of women into grassroots agricultural cooperatives and
   health and education programs by the diocese of San Cristobal de las Casas
   and a number of non governmental organizations (NGOs) . . . that began
   when the diocese adopted its preferential option for the poor in the 1970s
   . . . and deepened in the 1980s by projects initiated by university researchers, students, NGOs, and craft cooperatives located in San Cristobal.
3. The creation of the EZLN [Ejército Zapatista de Liberación Nacional] itself. Male-dominated community assemblies were transformed by
   women's demands for equal participation in the struggle. This was reflected in the Zapatista's Revolutionary Women's Law which states that
   all women should have the right to a life free of sexual and domestic violence, the right to choose one's partner and number of children, and
   the right to political participation on an equal footing with men.[28]

The processes at work that have aided indigenous women in asserting their
rights are contradictory, a combination of unintended aftereffects of colonization, the intervention of the Catholic church, and the growth of a pan-
indigenous movement in the Zapatistas. Our understanding of the liberative
effect of these three processes is not enhanced by beginning with an analysis
that starts with links between women and nature.

Rosemary Radford Ruether's observations concerning the differences between northern and southern ecofeminisms are appropriate here. Ruether claims that southern ecofeminists are primarily concerned with the "concrete reality of day-to-day life." They are also "less likely to idealize their own indigenous traditions, while still utilizing, appreciating and evaluating what is empowering and healing." And, they are "less likely to make blanket oppositions between all things Western and Christian as evil and all things from their own culture as good, or vice versa."[29] Although Chiapan women may not consider themselves ecofeminist, Ruether's insights are still appropriate. One might assume that *indigenous* equals *traditional lifeways.* That is not the case. Three main groups exist in highland Chiapas: traditionalists, Protestants, and Catholics (who generally follow liberation theology). Identity strongly corresponds to these categories. Conventional wisdom would suggest that traditionalists are the most "authentic" indigenous, reflecting greater continuity with their Mayan roots, shaped less by forces of globalization, with an ability to mount resistance to such forces given their religio-cultural resources. And there is indeed truth in these claims, although contradictions exist. An ecofeminist analysis might also assume that religions imposed by the colonizers—Roman Catholicism and evangelical Protestantism—would be more oppressive for women (and nature). Again, the situation is complicated.

Traditional women who have reached the end of childbearing increase their power and influence as shamans, midwives, grandmothers, and cargo holders. Yet, as I have already noted, symbolic and social constructs, many of which are rooted in nature, devalue women. Many women convert to Protestantism, viewing it as a means of bettering their lives. If their husbands quit drinking, as evangelical Protestants demand, household income generally increases and domestic abuse decreases.[30]

The Chiapan Catholic Church and liberation theologians, through numerous organizing efforts, also helped to create relatively autonomous political and social spaces for the growth of other non-governmental organizations (NGOs), including women's groups. The Diocesan Women's Committee, for example, made possible other women's organizations. In the report produced by the Bartolomé de Las Casas Human Rights Center concerning the 1997 massacre in Acteal, women informants claimed that they first learned of the concept of the rights of women in their catechism classes.[31]

Thus, Zapatista women may be traditionalists, evangelical Protestant, or Catholic. And from these divergent backgrounds and perspectives, they address gender discrimination within indigenous communities. Overemphasizing the West's economic system's culpability in harming indigenous women and nature, great as it may be, leaves us without analytic tools to uncover class, property, and gender relations within a particular nation, society, or group.

Women participants in the Zapatista movement challenge both the hegemonic project of the nation and global capitalism, as well as local structures of inequitable community power.

The preceding concern reflects the uneasiness some ecofeminists and ecofeminist critics share concerning the tendency to romanticize indigenous and other so-called Third World women. Mary Mellor notes a "tendency to treat all women in the south as having the same experience and potential as well as to romanticize their situations" which can result in a "new totalizing image of the valiant Third World woman . . . that deflects attention from divisions between women."[32] Salleh's quote which begins this essay reflects this tendency, as does her statement, "in the international division of labor, the domestic functions of indigenous and Third World women farmers are still bound up in care for earthly cycles."[33] And there is a logic to this "preference for indigenous cultures," which is, according to Sturgeon, "deeply implanted in ecofeminist theory." Sturgeon goes on to say:

> The ecofeminist critique of the hierarchical dualism of culture/nature at the heart of Western science and ideology therefore privileges those cultural and economic arrangements that are seen not to divide culture from nature, and that do not think of culture as superior to a degraded, inferior nature. This pervasive, and in many respects persuasive, critique of Western Enlightenment rationalism directs ecofeminists to non-Western cultures for examples of ecofeminist politics, culture, and economy. Further, in line with ecofeminist analyses of the interdependent relation between western culture/nature dualism and sexism, such "indigenous" cultures are seen as possible examples of more feminist societies. The term "indigenous" thus primarily signals for many white U.S. ecofeminists the extent that these cultures are nonindustrialized and therefore, from this perspective more ecological.[34]

Clearly practices and theories must account for the least advantaged, the victims of environmental degradation, unjust socioeconomic structures, and male domination. Emphasizing excluded knowledges and the agency of those women who fight for ecologically sustainable practices is important. Yet, several dangers exist as indigenous and Third World women are romanticized as the ultimate ecofeminists.

Fernando Mires begins his book, *El Discurso de la Indianidad: la cuestión indígena en América Latina*," by asking the question, "Who is an Indian?"[35] He concludes that "*indios*" to this day remain the products and imaginations of their "discoverers." The indigenous throughout history were repeatedly "discovered," first by negation (Columbus depicting Indians as an extension of nature), by death and genocide, by slavery, by evangelizing, by modern philosophy, and finally, by affirmation. The modern rediscovery of the indigenous, according to Mires, is done in the name of the defense of the *indio*. He writes, "the naturalizing of the Indian revives the image of the 'savage' with the difference that now it is framed positively."[36] In this modern version the indigenous is discovered by the state, in this case Mexico, as the bearer of and symbol for nationalism. Latin American leftists look to the indigenous to represent resistance against imperialism, whereas increased interest in ecology has led to a "rediscovery" of the *indios,* who are somehow more "natural" than the rest of us, for environmentalist purposes. And as Chiapan women who protest what they call "official indigenous discourse" tell us, in the process the real indigenous disappears, "assassinated by the functions they've been assigned within an ecosystem or by reactionary idealizations."[37]

Empirically, it may be difficult to sustain the notion of indigenous and Third World women's greater care for nature vis-à-vis men. Celia Nyamweru's essay in this volume about her fieldwork in the Coast Province of Kenya compares Mijikenda male and female practices and attitudes toward the *kaya* sacred forests, sites of cultural and spiritual significance. She finds that women's ways of using forest products are not more sustainable than men's. She writes, "If these Mijikenda were to sing their version of the Chipko women's song quoted by Shiva, 'What do the forests bear? Soil, water and pure air,' the choir might be dominated by men's voices." Women are also less likely to see sacred *kaya* forests as intrinsically valuable, and Nyamweru found no gender difference in the recognition of environmental problems and change.

This is not to say that women do not suffer disproportionately from environmental damage. Numerous studies demonstrate that this is the case. Rather, these cases caution us against constructing the "imaginary" indigenous or Third World woman. Championing a "symbolic indigeneity" is, according to Sturgeon, "ironically a form of antiracist discourse that . . . ends up despite good intentions, reconstituting white privilege. One way this occurs is through the racial essentialism of the idea of the indigenous, which erases all difference between and within the categories 'Native American' and 'Third World' and constitutes them as racialized Others to a white Self that is Western, modern, and industrialized."[38] As ecofeminists, we may conceptualize certain groups as being closer to nature, thus paradoxically replicating the dualism we repudiate. Claims made about indigenous and Third World women may actually serve to reassert patriarchal beliefs about women.

The demands from the indigenous women's movement in Chiapas are again helpful. The Women's Revolutionary Laws contain demands for the state as well as for their own communities. As Rosalva Aida Hernández Castillo writes, "These new voices question the dichotomy between tradition and modernity which official indigenism has reproduced which, in a certain way, the independent indigenous movement shares. . . . They are questioning the essentialist indigenous posture demanded of them which calls for defense of cultural traditions, questioning Mexican nationalism as well as autonomous indigenous discourse."[39]

Given these concerns, do I despair of the efficacy of ecofeminism? Not at all. I find great hope in ecofeminism's character as a social and political movement. Warren points out that ecofeminism "has always been a grassroots political movement motivated by pressing pragmatic concerns."[40] Sturgeon defines ecofeminism as an "oppositional political discourse and set of practices embedded in particular historical, material and political contexts."[41] If ecofeminism continues to become increasingly transnational, its theory will undoubtedly become further refined and address many of the concerns I raise in this essay. Salleh notes that ecofeminism is unique in its "transcultural sensibility and in being more than a single-issue identity politics."[42]

The fact that ecofeminist activists and thinkers of various stripes bring international attention to both environmental and feminist concerns and their convergence creates a political space for women in various parts of the world who otherwise may not have been heard. And, ironically, the idealization and romanticizing of indigenous and Third World women that I decried earlier also serves to increase visibility. If women from less affluent nations are epistemologically privileged within ecofeminism and considered the "experts," they well may, as is the case in Chiapas, contest idealizing tendencies of both official and ecofeminist discourse about the indigenous, the tribal, and the so-called Third World woman. This, in turn, makes for better ecofeminist theory and practice. As Sturgeon writes, "The discourse of indigeneity when coupled with claims about women's stake in environmentalism, which I have identified as a problematic element in U.S. ecofeminism in the late 1980s and early 1990s—as an *international political discourse* rather than a *theoretical* tool, opens up some possibilities."[43]

Sturgeon and Catriona Sandilands offer, I think, the most sophisticated accounts of how ecofeminism can serve as a democratizing political movement. For Sandilands, in spite of numerous theoretical problems grounded in what she sees as ecofeminism's standpoint epistemologies and identity politics, ecofeminism is fundamentally a democratizing movement. Key for Sandilands are coalitions that both privilege relations of solidarity and deny the primacy of any particular social location, thus speaking "to the proliferation of identities associated with new social movements."[44] Applauding Warren's call for epistemological diversity, Sandilands contends that even a "critique of essentialism ends up privileging a particular materialist way of knowing and thus is guilty of shutting down the epistemological diversity that the rejection of essentialism is supposed to produce."[45] Citing Carlassare, she demonstrates how the category woman, or indigenous, or Third World woman (etc.), may be "understood as a politically strategic invocation."[46]

Sandiland's analysis enables us to consider multiple "logics of domination" as part of a global movement or global coalitions:

> Any given site of struggle is not simply an obvious manifestation of a single logic of domination, thus subject to a predestined mode of political practice, rather, conditions and resistance are specific, and it is toward the empowerment of local communities to define and defend their particular interests that global politics are to be oriented . . . each struggle is thus impregnated with the meaning and desire of all the others . . . it is not that these disparate struggles have nothing to learn from one another or that there is no such thing as global capital, it is rather that the way in which each becomes an instance of the so-called globalization of the local serves as a form of retrospective symbolic over determination.[47]

Ecofeminism thus should live within the tension between the global and the local, the universal and particular, supporting the democratization of the global, while not reifying particular instances in forms of symbolic and theoretical overdetermination. Given the diverse resistance movements to the environmental damage caused by globalization, ecofeminism's ongoing negotiation of these tensions remains its hope and promise.

Let us return to the highlands of Chiapas. The case of Chiapas is not complicated at all. The Mexican military and paramilitary groups have committed atrocities against indigenous peoples, disproportionately women. Global capital displaces local peoples and colludes in the rapid destruction of the environment. This is deeply wrong. Yet, ecofeminism may not help us understand this particular case. Castillo writes, "These women aren't mere victims of patriarchal ideologies that hope to value their bodies to construct a Mexican nation or to perpetuate the native tradition. By calling themselves simultaneously Mexicans and indigenous these new voices propose to modify or change the characteristics of these 'imaginary communities.'"[48] As the women of the Chiapan Indigenous Women's National Congress say, they wish to *"cambiar permaneciendo y de permanecer cambiando"* (always change while staying the same and always remain the same while changing). In doing so, they hope to protect the land and preserve their culture while promoting their rights. They may or may not be ecofeminists. Yet, as an instance of the democratization of the global, their struggle concerns ecofeminists everywhere.

Notes

1. Ariel Salleh, *Ecofeminism as Politics: Nature, Marx and the Postmodern* (London and New York: Zed, 1997), 138, 104.

2. Lois Ann Lorentzen, "Reminiscing about a Sleepy Lake: Borderland Views of Place, Nature/Culture from a Salvadoran Context," in *Wild Ideas*, ed. David Rothenberg (Minneapolis: University of Minnesota Press, 1995).

3. Lois Ann Lorentzen, "Women, the Environment and Sustainable Development: Cases from Central America," in *Ecological Resistance Movements: The Global Emergence of Radical and Popular Environmentalism*, ed. Bron Taylor (Albany: State University of New York Press, 1995).

4. Lois Ann Lorentzen and Jennifer Turpin, eds., *The Women and War Reader* (New York: New York University Press, 1998).

5. Karen J. Warren, Introduction, in *Ecological Feminism*, ed. Karen Warren (London and New York: Routledge, 1994), 1.

6. Karen J. Warren, *Ecofeminist Philosophy: A Western Perspective on What It Is and Why It Matters* (Lanham, Md.: Rowman & Littlefield, 2000), 1–2.

7. Salleh, *Ecofeminism as Politics*, ix.

8. Salleh, *Ecofeminism as Politics*, 108.

9. Noël Sturgeon, *Ecofeminist Natures: Race, Gender, Feminist Theory and Political Action* (London and New York: Routledge, 1997), 23.

10. Warren, *Ecofeminist Philosophy*, 21.

11. Karen Warren, *Ecological Feminist Philosophies* (Bloomington: University of Indiana Press, 1996), 23.

12. Warren, *Ecological Feminist Philosophies*, x.

13. Warren, *Ecological Feminist Philosophies*, x.

14. Christine Eber, *Women and Alcohol in a Highland Maya Town: Water of Hope, Water of Sorrow* (Austin: University of Texas Press, 1995), 67.

15. Brenda Rosenbaum, *With Our Heads Bowed: The Dynamics of Gender in a Maya Community* (Albany: State University of New York Press, 1993), 69.

16. Rosenbaum, *With Our Heads Bowed*, 70–71.

17. Rosenbaum, *With Our Heads Bowed*, 33.

18. Janet Biehl, *Rethinking Ecofeminist Politics* (Boston: South End Press, 1991), 51.

19. Irene Silverblatt, *Moon, Sun, and Witches: Gender Ideologies and Class in Inca and Colonial Peru* (Princeton, N.J.: Princeton University Press, 1987).

20. Mercedes Olivera, "La compra de la novia, símbolo de subordinación servil en los albores del siglo XXI," *Cuadernos feministas* 1, 2 (1997): 19.

21. Bina Agarwal, "The Gender and Environment Debate: Lessons from India," *Feminist Studies* 18, 1 (1992): 3–19.

22. Warren, *Ecological Feminist Philosophies*, xiv.

23. Val Plumwood, "The Ecopolitics Debate and the Politics of Nature," in *Ecological Feminism*, ed. Karen J. Warren (London and New York: Routledge, 1994), 70.

24. Warren, *Ecofeminist Philosophy*, 2.

25. Rosenbaum, *Heads Bowed*, 33–36.

26. Eber, *Women and Alcohol*, 27.

27. Eber, *Women and Alcohol*, 47.

28. Neil Harvey, *The Chiapas Rebellion: The Struggle for Land and Democracy* (Durham, N.C.: Duke University Press, 1998), 223–24.

29. Rosemary Radford Ruether, *Women Healing Earth: Third World Women on Ecology, Feminism, and Religion*, ed. Rosemary Radford Ruether (Maryknoll, N.Y.: Orbis Books, 1996), 6–7.

30. Eber, *Women and Alcohol*, 245.

31. Centro de Derechos Humanos Fray Bartolomé de las Casas, "Antes y después de Acteal: voces, memorias y experiencias desde las mujeres de San Pedro Chenalhó," in *La Otra Palabra: mujeres y violencia en Chiapas, antes y después de Acteal*, ed. Rosalva Aida Hernández Castillo (San Cristóbal de Las Casas, Chiapas, México: Centro de Investigaciones y Estudios Superiores, en Colectivo de Encuentro entre Mujeres [COLEM] y el Centro de Investigación y Acción para la Mujer [CIAM], 1997).

32. Mary Mellor, *Feminism and Ecology* (New York: New York University Press, 1997), 35.

33. Salleh, *Ecofeminism as Politics*, 139.

34. Sturgeon, *Ecofeminist Natures*, 114.

35. Fernando Mires, *El Discurso de la Indianidad: la cuestión indígena en América Latina* (San Jose, Costa Rica: Editorial Departamento Ecuménico de Investigaciones, 1991), 11.

36. Mires, *El Discurso*, 37.

37. Mires, *El Discurso*, 37.

38. Sturgeon, *Ecofeminist Natures*, 113.

39. Rosalva Aida Hernández Castillo, *La Otra Palabra: mujeres y violencia en Chiapas, antes y después de Acteal*, ed. Rosalva Aida Hernández Castillo (San Cristóbal de Las Casas, Chiapas, México: Centro de Investigaciones y Estudios Superiores, en Colectivo de Encuentro entre Mujeres [COLEM] y el Centro de Investigación y Acción para la Mujer [CIAM], 1997), 133–34.

40. Warren, *Ecofeminist Philosophy*, 34.

41. Sturgeon, *Ecofeminist Natures*, 3.

42. Salleh, *Ecofeminism as Politics*, 110.

43. Sturgeon, *Ecofeminist Natures*, 139.

44. Catriona Sandilands, *The Good-Natured Feminist: Ecofeminism and the Quest for Democracy* (Minneapolis: University of Minnesota Press, 1999), 98.

45. Sandilands, *Good-Natured Feminist*, 113.

46. Sandilands, *Good-Natured Feminist*, 114.

47. Sandilands, *Good-Natured Feminist*, 132.

48. Hernández Castillo, *La Otra Palabra*, 134.

5

Traditions of Prudence Lost: A Tragic World of Broken Relationships

Aruna Gnanadason

Introduction

Aruna Gnanadason, who works on gender concerns at the World Council of Churches, reflects on how globalization is shattering centuries-old Indian agricultural traditions. Aruna sees globalization as a form of violence against millions of people and an intentional death to the land. She also shows how the Dalit and indigenous women, the lowest castes of the Indian caste system, bear the greatest impact of the destruction of creation and have the least access to resources. In addition, the ideology of the caste system supports the belief that the Dalit women pollute the waters for the upper castes. Aruna demonstrates the importance of culture, race, and class in conversations about ecofeminism and globalization.

MUNIYAMMA INHERITED AN ACRE of ancestral land. This she tilled and cared for with tenderness. It was not enough to sustain her and her family of five, but it fed them with the basic foods they needed. The land was sacred to her, having belonged to her family for many generations. It had been gifted to her at the time of her marriage. Nagappa, her husband, worked as an office attendant in a nearby town to augment their income. Life was difficult, but fulfilling. They were one of the few Dalit families who owned land in Pattandur village, some fifteen kilometers from Bangalore, the metropolitan city. (The Indian social structure is built on the caste system of graded subjugation. There are four main caste groups: the Brahmins, or priestly caste, at the top, followed by the warrior caste, then the merchant class and, at the bottom, the *shudras*, the working classes. The Dalits are outside this structure and are considered unclean and polluting. For centuries they have been treated as untouchables, and today they continue to face discrimination and violence.) For Muniyamma every new season of preparing the land for sowing was filled with ceremony. The seeds that had been saved from the previous harvest were brought out and, amid rituals and prayers, the sowing was done. Rituals accompanied each stage of the growth of the grains. Harvest was a particularly auspicious time with great celebrations, when the whole village came together for food and drink.

Then it all fell to pieces: the liberalization of the market hit Bangalore. The city was hailed as the "Silicon Valley of the East." The advertisers proudly announced, "Rome is the city of the past and Bangalore the city of the future." An international tech park grew as a free trade zone near Muniyamma's village. Land became dear, and the villagers were offered large sums of money to sell their lands to the big telecom industrial expansionists. Muniyamma resisted as long as she could, but then the pressure told on her and she caved. Her alcoholic husband did not help the situation—as a mere woman in that patriarchal household her voice of reason did not hold much weight. Muniyamma and her family moved into the nearby town, and everything has changed. They live in a small house, in a settlement for the urban poor, with no garden, no extended family, and, most importantly—no land! Muniyamma feels in her heart the broken relationship with her land—she lost a part of her soul when she lost her land.

Bangalore, this great "city of the future," has been turned upside down. Roads and roads and more roads are in preparation to service the industrial sector, to serve free trade. For the rich, all this is progress, all this means comfort and confidence. But, as the car we were traveling in made its very slow progress on a road almost fully dug up, I watched a family of children, possibly of a construction worker, on that mighty road that was growing above their heads. In a makeshift residence, I saw a child of about ten, apparently left in charge of her younger brother of about six and a sister of about three. They were squabbling over a piece of bread, perhaps that their mother had left for them. I could not but ask myself, "Whose earth, whose future?" What of Muniyamma and this family of children, who will never enjoy the fruits of this city's so-acclaimed "progress"? I asked myself, "Whose India?"

In the World Trade Organization (WTO) talks in Seattle, India did all it could to continue to embrace free trade and globalization. The Indian delegation went to Seattle to discuss further trade liberalization. They went there determined to "emulate the West," as the *Economist* describes their intentions.[1] They came away disillusioned that "labor rules" were being imposed on trade agreements. This, the delegation was unhappy about. All else, including the liberalization of agriculture and of textiles, they were ready to negotiate. The discussions did not include Muniyamma and the millions of women like her who bear the brunt of these agreements. The state assumes monopoly over the natural resources and, with it, the right to allow and direct the flow of industry, international or national. They do what they will with the land. Ironically, India is still basically an agrarian society, and now with the lift of all controls on the import of agricultural products, the Indian consumer will be soon eating cheap imported rice. What of the life and livelihood of the farming community in India? And more seriously, how is this community going to survive when its intricate web of relationship with the earth is broken? Liberalization of the economy and with it privatization is taking over sector after sector of life and livelihood—no one seems to care where the nearly one billion people will find work or sustenance. The encroachment into creation is stark!

Every year, when our family visits our ancestral village, Thittuvillai, we see more changes, more takeover of Mother Earth by those who have the power and money to do so. Thittuvillai is a small village in the southernmost district of India. Some ten years ago it was still a little piece of heaven on earth. Rich greenery—paddy fields and the verdant mountain range of the Eastern Ghats—surround this little village. It is a region that has thrived for centuries and is steeped in mythology and nature-affirming folk tales. Thadagathi is one of these mountains: she is a slain "demon queen," according to a popular version of the story, killed by the Lord Rama, one of the leading figures in the epic *Ramayana.*

Thadagathi was the reigning queen of the region. Dark skinned and "ugly," according to the standards of the invading Aryan culture, she became a "demon" in their eyes. The story goes that Rama was hunting in the region, on his way to Lanka to save his abducted queen Sita from the hands of her abductor the King of Lanka, Ravana. Thadhagathi challenged Rama and asked him to desist from killing the wildlife of the territory over which she ruled. Rama was not used to having his authority challenged by anyone, let alone by a dark-skinned "ugly" woman, and so he killed her. He is said to have deeply regretted having killed a woman, but he had his way. Read from the perspective of the "small people" the story gets a whole new interpretation. She undoubtedly represents the Dalit peoples of India. The story reflects Dalit culture that has built into it principles of respect and a protective attitude to the earth and the things of the earth. Thadagathi lies there—a mountain, serene and majestical, having lost out to the selfish greed of the powerful one, who considered hunting to be his birthright. Thadagathi lies there dead, reminding India of all that we are losing! The Eastern Ghats are, year after year, losing their greenery. A region, which has for centuries survived with its mountain rivers and lakes, blessed in the protective care of Thadagathi, is now slowly dying. The water is being channeled into the neighboring district to water the Koodankulam nuclear reactor and other industrial projects. The paddy fields are slowly disappearing as the struggling farming community is forced to sell out to those who have the money to build bigger homes. It is one more symbol of the dying India.

Arundathi Roy, the writer, has joined the protest against the dam projects on the River Narmada, giving public support to the long struggle of the indigenous peoples for their life and livelihood. She draws attention to how the destruction of peoples, their livelihood, their cultures, their lives, is callous, brutal, and violent. Ram Bai, whose village was submerged when the Bargi dam was built on the Narmada, said, "Why didn't they just poison us? Then we wouldn't have to live in this shit-hole and the Government could have survived alone with its precious dam all to itself."[2] She now lives in a slum in Jabalpur. Roy draws attention to the plight of the displaced people who, moved off their land, now earn a few rupees a day to stay alive, instead of getting all that they need from the forest—food, fodder, fuel, rope, gum, tobacco, medicinal herbs, housing materials, even tooth powder: "Instead of a river, they have a hand pump. In their old villages, they had no money, but they were insured. If the rains failed, they had the forest to turn to. The river to fish in. Their livestock was their fixed deposit. Without all this they are a heartbeat away from destitution."[3]

In recent years, the River Narmada dams project is the most dramatic expression of the kind of violent "development" path India walks on. The River Narmada flows for 1,300 kilometers through three Indian states—Madhya Pradesh, Maharashtra, and Gujarat—before it empties itself into the Arabian Sea. The project includes plans to build 3,200 dams on this river and its forty-one tributaries. Thirty will be major dams, one hundred thirty-five medium, and the rest small. It will alter the ecology of this valley; it will affect the lives of twenty-five million people and submerge four thousand square kilometers of natural deciduous forest. Twenty-five million people who have lived in this river valley, who have been intricately interwoven with the ecosystem and with each other by an ancient web of interdependence, will be, in one cool sweep, pushed out of existence.

All this is indeed violence against half a million people. It is intentional death to the land. It is an intentional way to break a people's sacred bond with the land. In the words of Arundathi Roy:

> To slow a beast, you break its limbs. To slow a nation, you break its people. You rob them of volition. You demonstrate your absolute command over their destiny. You make it clear that it ultimately falls to you to decide who lives, who dies, who prospers, who doesn't. To exhibit your capability you show off all you can do, and how easily you can do it. How easily you could press a button and annihilate the earth. How you can start a war, or sue for peace. How you can snatch a river away from one and gift it to another. How you can green a desert or fell forest and plant one somewhere else. You use caprice to fracture a people's faith in ancient things—earth, forest, water, air.[4]

But there is hope yet. The Narmada struggle has demonstrated the great resilience of the people to sustain a struggle. Even as I write this article, the fight to protect the land and life of the indigenous peoples from the dam projects continues. Every now and then people, largely indigenous people, gather at one of the dam sites to protest. They camp there and say that they will not move even if the waters are opened on to them. They would be submerged and perish along with their land!

All over the world there is now a greater consciousness about the environment and the urgency to protect it. Just a decade ago governments, particularly in the South, and even social movements and leftist political parties tended to dismiss this issue as irrelevant, but today the newspapers and other media constantly project the dimensions of the problem. The state has finally taken upon itself a greater responsibility to act. One of the Indian states, for instance, has banned the use of plastic covers for packing purposes in shops and other establishments as of January 2000. These changes are good, although it may be too late, and the actions of the government are not above suspicion, if we see the state of the country's environment. More importantly, does banning plastic covers get to the root of the problem? Roy quotes Larry Rasmussen as he reminds us that "the debt to nature cannot be paid person-by-person in recycled bottles or ecologically sound habits, but 'only in the ancient coin of social justice.'"[5]

Bina Agarwal lays out how, according to government of India statistics, 19.5 percent of the country's geo-area was forested and this area is declining at an estimated 1.3 million hectares a year. Again, according to official estimates, in 1980 56.6 percent of the land suffered from environmental problems, especially water and wind erosion. In some canal projects one-half of the area is so waterlogged that it cannot be irrigated and cultivated. The land area under periodic floods doubled between 1971 and 1981. The soil is losing its fertility due to excessive use of chemical fertilizers. The level of ground and surface water is falling due to indiscriminate sinking of tube wells. Fertilizer and pesticide runoffs into natural water sources have destroyed fish life and have polluted water for human use in several areas.[6]

The commissioner of traffic of the city of Chennai, in a recent TV interview, said that the car population in the city has increased from 700,000 cars in 1994 to 1,100,000 in the beginning of 2000. And what does the government of India do? It agrees, in WTO negotiations, to end import controls on used cars! For some time India had insisted at least on quantitative controls—but having lost out to the United States in the WTO's Dispute Settlement process, India was forced to concede to the lifting of all controls. The fall from a position of strength to a position of total surrender has been rapid. Just some ten years ago, India boasted about the growth of her own national car industry. For decades we were quite satisfied with just two or so car models. Then slowly we succumbed to the penetration of foreign companies. You name the brand, we have them all—Ford, Peugeot, Fiat, Suzuki, Hyundai, Volvo, and Mercedes Benz! And now, why not clutter the overpolluted, accident-prone city roads with cars used and discarded in the United States, Europe, and Japan? Ironically, at the same time we are told that there is already a glut in the car trade. Who cares if the earth groans and weeps at the insanity of it all?

This is the complex reality in which we need to find a response to the difficulties and injustices faced by women like Muniyamma and Ram Bai, struggles that go back to an ancient past represented by stories such as that of Thadagathi. There is no easy analysis—traditional questions are inadequate and quick or simplistic answers to those questions will not suffice. The response has to be multifaceted and inclusive of the dreams and hopes of those most marginalized, those most affected by policies and programs that disrespect the earth and its people. In thinking about a response or responses, there are several important aspects of the discussion on an ecofeminist view from India, and from most of the Third World, that need to be kept in mind.

There is a need to recognize that there are other layers of injustice that aggravate the situation. It is not possible to speak of "women" as one oppressed category. Any feminist vision of creation has to embrace and acknowledge cultural and social forms of discrimination that make the lives of some women even more precarious than other women. Indian society is inherently unjust and is built on the system of graded subjugation of caste and religious identity. At the very bottom of the cultural scale are Dalit and indigenous women, who bear the greatest impact of the destruction of creation and the resources of the earth. For example, this inherently unjust society has ensured that Dalit women have even less access to water. They are said to pollute the water of the upper caste wells and therefore they have to depend on the only well on their side of the village. Its drying up could mean an endless wait for their vessels to be filled by upper-caste women or long walks away from home in search of water.

Therefore, we need to emphasize that it is not possible, in India and in most of the Third World, to posit women as a unitary category; we need to differentiate among women by class, race, caste, ethnicity, and so on. It is necessary to critique ecofeminist discourse, largely generated in the northern hemisphere, that "ignores forms of domination other than gender, which also impinge critically on women's position."[7] Women's lived relationships with nature are influenced and are different according to different local contexts, not just between continents or worlds, but within nations. It is not possible to speak, as some ecofeminists do, of a female "essence," which is unchangeable and irreducible. "Equally, it is critical to examine the underlying basis of women's relationship with the non-human world at levels other than ideology (such as the work women and men do and the gender division of property and power) and to address how the material realities in which women of different classes (castes/races) are rooted might affect their responses to environmental degradation."[8] The plea is to analyze more deeply why different communities of women get involved in the struggles to protect the earth. All such actions are noble and to be affirmed, but each is related to a specific experience of relationship with the things of the earth. Therefore, ecofeminist analysis must acknowledge the internal contradictions among women and cannot ignore that there are systems and structures in place that distribute power over the use of the resources of the earth unjustly and unequally, even among women. Agarwal[9] and Gabriele Dietrich[10] both express concern that Vandana Shiva does not take into account the connection between patriarchy and caste. She depends on resources only from a Brahmanical philosophical heritage to challenge the destruction of the earth, and not from the points of view of Dalits and indigenous peoples. There is a point in seeing that the resources to protect the earth come from those who are closest to the earth and whose survival depends on its protection.

That is not to fall into the erroneous notion of competition, as Lois K. Daly does when she uses the unfortunate expression "competing feminisms"[11] to describe this diversity. There are no competing discussions between various feminist worldviews; there are just many entry points and perspectives in feminist discourse. It would indeed be counterproductive if we, as women, keep ourselves divided on ideological lines. We should in humility offer our many experiences to each other and learn from each other, bringing our plurality of visions to a common commitment to affirm life. There are no "competing feminisms," only a wide range of experiential bases that inform our theological visions. In a world that so desperately needs resources to counter globalizing trends, I believe we need to find ways to offer all our gifts at a common table so that together, as women, we can discover ways forward to offset the destructive trends in our world.

Ironically, the promise of abundant life is an elusive dream for millions of women in the world—they can barely survive. Therefore my own entry point into ecofeminist discourse is the life of women who are daily engaged "in the production of survival,"[12] as Vandana Shiva, the feminist environmentalist, describes it. I draw my inspiration from the many ways in which women find spiritual resources for their struggle. Therefore, my ecofeminist vision is not some romantic or esoteric vision; it is based on a plea for sanity; it is a cry that we recognize as sin the destruction of the earth. By this I mean all that is on this earth, human and otherwise. To reverse this madness of destruction is indeed an urgent imperative for our times. Therefore, it is valid to speak of "the earth community" rather than "the environment." I want to stress that for many Third World feminists, when we speak of the survival of trees, of the air, of the land and seas, such a concern is inextricably linked with the improvement of the quality of life, indeed, with the survival, of all people, particularly women, who bear the most severe consequences from the degradation of the earth.

Women in India, as in many other parts of the Third World, depend on the earth for their survival. Sex role divisions of work ensure that women do the most strenuous kinds of work *in close proximity to the resources of the earth—*gathering food and fuel and collecting water. Women are expected to care for the families, often singlehandedly. When there is a depletion of resources, women need to go even further in search of food, water, firewood, and means of livelihood. With the depleting resources, women's work and working hours increase. Firewood is still the single most important source of energy in India, in many places 65 percent of domestic energy.[13] In most places women need to collect firewood from the forest and the commons; it is not purchased. The same goes for fodder. As a woman in northwest India puts it: "When we were young, we used to go to the forest early in the morning without eating anything. There we would eat plenty of berries and wild fruits . . . drink the cold sweet water of the *banj* (oak) roots. . . . In a short while, we would gather all the fodder and firewood we needed, rest under the shade of some huge tree and then go home. Now, with the going of the trees, everything else has gone too."[14] It is women who suffer the most when drinking water becomes scarce. Millions of women, in India and in many parts of the world, spend endless hours walking miles in search of water every single day of their lives. But environmental degradation has even more serious consequences in the lives of women. In one state in north India, according to an activist, "the growing hardship of young women's lives with ecological degradation has led to an increased number of suicides among them in recent years. Their inability to obtain adequate quantities of water, fodder and fuel causes tensions with their mothers-in-law (in whose youth forests were plentiful) and soil erosion has compounded the difficulty of producing enough grain for subsistence in a region of high male outmigration."[15]

The United Nations Development Program's (UNDP) 1999 *Human Development Report* once again reminds us that the world we live in is both intentionally unjust and inherently women-unfriendly. India is ranked in the 570th spot in the Human Development Index (HDI).[16] Canada is in the first place. The index is a very helpful tool. It starts from the premise that the real wealth of a nation is its people and includes, not just the usual economic yardsticks, but such indicators as health, education, gender, technological progress, and the like. There has been a fall in India's HDI ranking between 1985 and 1997, in spite of the opening of markets and the liberalizing of foreign investments, supposedly in a bid to improve the quality of life of its people. The UNDP reports that 19 percent of India's over one billion people still do not have access to clean drinking water and 25 percent do not have access to health care and 71 percent to sanitation. While Canada's maternal mortality rate is six maternal deaths per 100,000 live births, India's is 570 per 100,000! In connection with the gender-related development index, India again is dismally low, ranking 112th. Canada, again, stood in first place, but even Saudi Arabia and the conflict-ridden countries of Indonesia and Sri Lanka scored higher than India.[17] All this to underline the fact that any reflection on the environment has to start from the struggles for survival of women. This is imperative because this is the reality for the majority of women in the world.

Based on research in a region in central India, M. Gadgil and K. C. Malhotra point out that castes within Indian society, particularly pastoralists and nomads that directly depended on natural plant and animal resources, developed specific ways of utilizing these resources, which, coupled with territoriality, ensured that a particular limiting resource in a particular geographical region was more or less exclusively utilized by a particular lineage. The lineage would be aware that the resource had supported it for generations past and will have to continue to support it for generations to come. . . . The result of these practices would be *to promote the evolution of cultural traditions of prudent exploitation of the natural resources.*"[18]

They go on to underline that traditions of prudence depend on the condition that some other lineage does not usurp the resource when it becomes available at a later time, and the resource should continue to be of value to the lineage adopting prudence. "The mode of resource utilization evolved by the Indian society clearly fulfils these conditions. We therefore, expect the evolution of a number of cultural practices resulting in a sustainable use of natural resources by the caste groups that constitute not only the genetic but also the cultural units of the Indian society."[19] Several specific cases are cited to substantiate this. The Dheevar caste of Bhandara district of Maharashtra never catch fish going upstream on spawning migration, although the fish are exhausted and easy to catch; there are entire sacred groves and ponds in which no plant or animal is damaged and some species of plants and animals survive only in such protected localities; and monkeys, peafowl, banyan and fig trees, and a variety of plants and animals are regarded as sacred and are protected widely in many parts of India.[20]

It is the tradition of prudence that is at the heart of the well-known Chipko movement that Vandana Shiva analyzes so excellently. She argues that the women in the Gharwal region have a special dependence on the earth and a special knowledge of the earth. The largely women-centered Chipko movement used the only forms of resistance available to them; they clung to the trees defying the saws of the contractors. They had also protested mining operations in the region. They were simply protecting the lives of their communities—in 1975 there had been landslides that threatened the existence of three hundred villages. In Shiva's words:

> Bachni Devi of Adwani led a resistance against her own husband who had obtained a local contract to fell the forest. The forest officials arrived to browbeat and intimidate the women and Chipko activists, but found the women holding up lighted lanterns in broad daylight. Puzzled, the forester asked them their intention. The women replied, "We have come to teach you forestry." He retorted, "You foolish women, how can you who prevent the felling know the value of the forest? Do you know what forests bear? They produce profit and resin and timber." And the women immediately sang back in chorus:
> "What do the forests bear?
> Soil, water and pure air
> Soil, water and pure air
> Sustain the earth and all she bears."[21]

These peasant women challenged "the reductionist commercial forestry system on the one hand and the local men who had been colonized by that system, cognitively, economically and politically on the other."[22] The traditional knowledge of the women who know what happens to their communities when the soil, air, and water are polluted is not taken cognizance of in the minds of planners and forest officials. The intricate and tender bond between the communities and creation is broken—and the foolish ex-

ploitation of the forests goes on unabated. An elder from one of the indigenous nations in India once told me that the annual flooding and destruction by the angry Thungabadra river can be explained as God's knuck-
les knocking on our heads, because we continue to not know how to
behave!

India's inherently unjust social and patriarchal structure has clearly contributed to the breakdown of the sustainable use of the resources of the
earth, but British rule also contributed: "The British imposed much higher
levels of demands on natural resources. . . . They took over as Government
property vast resources which, until then, were owned communally. These
resources were then rapidly depleted through commercial exploitation, a
trend that has accelerated over the last three decades since independence."[23]
Vandana Shiva talks of how women's impoverishment increased during
colonial rule:

> Those rulers who had for centuries subjugated and reduced their own women to
> the status of de-skilled, de-intellectualized appendages, discriminated against the
> women of the colonies on access to land, technology and employment. The eco-
> nomic and political processes of colonial underdevelopment were clear manifes-
> tations of modern western patriarchy, and . . . women tended to be the greater
> losers. The privatization of land for revenue generation affected women more
> seriously, eroding their traditional land use rights. The expansion of cash crops
> undermined food production, and when men migrated or were conscripted into
> forced labor by the colonizers, women were often left with meager resources to
> feed and care for their families.[24]

The British established state monopoly over the forests and curtailed the cus-
tomary rights of local populations to these resources. The colonial power also
introduced the concept of "scientific" forest management and the practice of
encouraging commercially profitable species. In addition, there was the inva-
sion of the forests for exploitation by private European and Indian contractors, especially for building railways.[25] With the exploitation, the sacred
groves, the sacred mountains, the sacred trees, and the sacred rivers were all
desecrated. Economic profit became the value placed on them.

A model of "development" promoted on the lines of Western science and
technology as a direct consequence of the industrial revolution in Europe cannot
be appropriate for an agrarian economy such as India. This industrial model of
development has become all pervasive. The entire world was forced to model itself on the lines of the colonizing west—which had gone through a different history and therefore a different process of development. Europe did not experience
the subjugation and exploitation that colonization entailed. It was assumed that
"western style industrial development" was possible for all, even those countries

that were essentially agricultural, which now had to transform their economies to keep pace with industrial expansion:

> Concepts and categories about economic development and natural resource utilization that had emerged in the specific context of industrialization and capitalist growth in a centre of colonial power, were raised to the level of universal assumptions and applicability in the entirely different context of basic needs satisfaction for the people of the newly independent Third World countries. Yet as Rosa Luxembourg has pointed out (in *The Accumulation of Capital*, London: Routledge and Kegan Paul, 1951) early industrial development in western Europe necessitated occupation of the colonies by the colonial powers and the destruction of the "local (natural) economy."[26]

This concept of "development" was clearly based on capital accumulation and commercialization for the generation of profits. This implied not only the creation of wealth *but* also the creation of poverty and dispossession. This Eurocentric (and later America and Japan centered) model of development legitimized colonialism and imperialism and the economic choking and neo-colonialism that took over nation after nation in the South. This submerged all other civilizations, all other cultures, all other historical experiences. It ignored highly developed systems of philosophical and religious thought and asserted that the Western paradigm was a so-called civilizing force in a supposedly uncivilized world.

Along with "statization of communal lands" mentioned earlier, Bina Agarwal identifies other outcomes of the present development paradigm that have contributed to the present state of the environment. She speaks of the process of privatization by those who could afford to buy land, but also the policy of "redistribution" of common lands supposedly for the poor, but which finally benefited rich farmers. There is also the erosion of community resource management systems (the loss of the tradition of prudence referred to earlier). Population growth needs also to be mentioned here, although the discussion has to be nuanced by its link with poverty and the low status of women. Additionally, there is the question of the imposition of agricultural technology and the erosion of local knowledge systems.[27] What the colonialists started has been strengthened and continued by successive Indian governments.

A new threat to creation is in the area of genetic engineering and intellectual property rights on seed varieties. This control of the regenerative capacity of the earth has had the most devastating effect on the Indian farming community and particularly on Indian women such as Muniyamma. "The land, the forests, the rivers, the oceans, the atmosphere have all been colonised, eroded and polluted. Capital now has to look for new colonies to invade and exploit for its further accumulation. These new colonies are . . . the interior spaces of the bodies of women, plants and animals."[28] What is autonomous, free, and self-generative is rapidly being brought under the control of technology and industry for profit. All this is ironically termed as progress. "Violence and plunder as instruments of wealth creation do not just belong to the history of colonisation . . . they are essential for the colonisation of nature and of our bodies through the new technologies. As before, those who are exploited become the criminals, those who exploit require protection. The North must be protected from the South so that it can continue its uninterrupted theft of the Third World's genetic diversity."[29]

The World Trade Organization has drawn up the most far-reaching multilateral agreement on intellectual property rights called TRIPS (Trade-Related Aspects of Intellectual Property Rights, 1995). Industrial countries hold 97 percent of all patents worldwide. In 1995, more than half of global royalties and licensing fees were paid to the United States. The use of intellectual property rights is alien to many developing countries. More than 80 percent of the patents that have been granted in developing countries belong to residents of industrial countries. And so, indigenous plant varieties and seeds are appropriated and claimed by others and patented. The laws ignore cultural diversity and diversity of views on what can and should be owned, from plant varieties to human life. The result is a silent transfer (theft) of centuries of knowledge from developing to developed world. Even the UNDP would acknowledge that "local plant breeding is essential for adapting seeds to the ecosystem and maintaining bio-diversity. The 1.4 billion rural people relying on farm-saved seed could see their interests marginalized."[30]

It is important that this complex reality is kept in mind when we seek solutions and offer an alternative vision and response. While I have focused on India as a case, there is much that fits into the framework of other Third World contexts and the life of Third World women. There is a different worldview, a different analysis, and definitely different resources, proper to Third World women that they use to survive. Bina Agarwal wonders whether there is a need to use an alternative formulation to describe this experience. She suggests the term "feminist environmentalism" to the often-used "ecofeminism."[31] In the final analysis, it is not the labeling of different contributions that matters—what matters is that it is understood that something new can be offered from this specific experience. In India, as in Kenya and in Brazil, it is poor women who have been most affected by environmental destruction, and it is they who are most active in the movements to protect the earth. It is the survival of these women that is the most precarious.

So in a broken world, the most pertinent question relates to the sources of prudence. In India and in most of the Third World it is the traditional knowledge of prudence that has been silenced. As a Christian, I look for resources in my own faith heritage and recognize the need to look to the wisdom tradition and to the prophets who spoke sharply regarding the loss of the tradition of prudence. Amos the prophet, in chapter 5, verses 11 through 15, reminds us that "the prudent will keep silent in such a time; for it is an evil time." In a text rich with imagery, he calls on the people "to hate evil and love good and establish justice at the gate." We need to recover these sources of hope in a tragic world of broken relationships. Let us listen to the earth and the children of the earth.

Notes

1. *The Economist* 353, 8149 (December 11, 1999): 13.

2. Arundathi Roy, *The Greater Common Good* (Bombay: India Book Distributors Ltd., 1999), 6.

3. Arundathi Roy, *Greater Good*, 38.

4. Arundathi Roy, *Greater Good*, 60–61.

5. Arundathi Roy, *Greater Good*, 142.

6. Bina Agarwal, "The Gender and Environment Debate," in *Gender and Politics in India*, ed. Menon Nivedita (Delhi: Oxford University Press, 1999), 109–10.

7. Bina Agarwal, "Gender Debate," 100.

8. Bina Agarwal, "Gender Debate," 101.

9. Bina Agarwal, "Gender Debate."

10. Gabriele Dietrich, "Women, Ecology and Culture," in *Gender and Politics in India*, ed. Menon Nivedita (Delhi: Oxford University Press, 1999), 72–95.

11. Lois K. Daly, "Ecofeminism, Reverence for Life and Feminist Theological Ethics," in *Liberating Life: Contemporary Approaches to Ecological Theology*, ed. Charles Birch, William Eakin, and Jay B. McDaniel (Maryknoll, N.Y.: Orbis, 1990), 93.

12. Vandana Shiva, *Staying Alive: Women, Ecology and Development* (London: Zed, 1989).

13. Bina Agarwal, "Gender Debate," 119.

14. Bina Agarwal, "Gender Debate," 119.

15. Bina Agarwal, "Gender Debate," 121.

16. United Nations Development Program, *Human Development Report*, 1999.

17. United Nations Development Program, *Human Development Report*, 1999.

18. M. Gadgil and K. C. Malhotra, "The Ecological Significance of Caste," in *Social Ecology*, ed. Ramachandra Guha (Delhi: Oxford University Press, 1998), 36.

19. Gadgil and Malhotra, "Significance of Caste," 36.

20. Gadgil and Malhotra, "Significance of Caste," 37.

21. Vandana Shiva, *Staying Alive*, 74.

22. Vandana Shiva, *Staying Alive*, 74.

23. Gadgil and Malhotra, "Significance of Caste," 37.

24. Vandana Shiva, "The Impoverishment of the Environment: Women anᴗ dren Last," in *Ecofeminism*, ed. Maria Mies and Vandana Shiva (London: Zed, 1993), 72.

25. Bina Agarwal, "Gender Debate," 110.

26. Vandana Shiva, *Staying Alive*, 1.

27. Bina Agarwal, "Gender Debate," 110–17.

28. Vandana Shiva, "The Seed and the Earth: Biotechnology and the Colonisation of Regeneration," in *Minding Our Lives: Women from the South and North Reconnect Ecology and Health*, ed. Vandana Shiva (Delhi: Kali for Women, 1993), 129.

29. Vandana Shiva, "The Seed," 136–37.

30. United Nations Development Program, *Human Development Report*, 1999, 67–68.

31. Bina Agarwal, "Gender Debate," 96.

III

REGIONAL AND TRANSNATIONAL EXPRESSIONS OF ECOFEMINISM AND RESPONSES TO GLOBALIZATION

T HE ESSAYS IN THIS SECTION SHOW the many ways ecofeminism intersects with cultures and religions. They demonstrate how ecofeminism can be influential in political movements, global environmentalism, and development agendas. These essays make us aware that one cannot speak of *ecofeminism* but of *ecofeminisms* and that ecofeminism(s) can be both theoretical and pragmatic tools.

6

Ecofeminist Natures and Transnational Environmental Politics

Noël Sturgeon

Introduction

Noël Sturgeon, an ecofeminist sociologist, provides a well-documented analysis of how different strands of ecofeminism have been used in international development. She shows how essentialism in some ecofeminisms, while often criticized, can be used as a strategic tool for change. Noël describes in detail how ecofeminist theories are used in different parts of the world. She analyzes ecofeminist influence on the women and development agenda, as well as theoretical and pragmatic problems.

TENSION EXISTS BETWEEN ECOFEMINIST definitions of diversity that privilege differences based on U.S. racial categories, and notions of diversity based on "international" difference. In this essay, I decenter the U.S. context in order to consider the deployment of ecofeminist conceptions of race and gender within a transnational context. I want to ask about their political results within a particular historical, disciplinary, and political context in the early 1990s in order to explore the conditions under which "strategic essentialisms" operate, and to generate ways of assessing their effects. I start by sketching two interrelated contexts. One is the field of development studies, which, from 1970 to 1990, had experienced shifts from "development" to "women in development" to "women, environment, and development." The second context is a phenomenon I call the "globalization of environmentalism," or the hegemonic contests over the meaning and use of "environmentalism" within a post–cold war transnational political arena. Finally, I look at a specific example of the development of an implicitly ecofeminist discourse as a mobilizing tool by an organization called WEDO, or Women's Environment and Development Organization, which was founded in 1990 to orchestrate a "women's voice" within UN deliberations over the intersection between environment and development. What I want to show here is the way in which "ecofeminism," rather than being a fixed group of movement actors or organizations, or even a set of circumscribed theories or analyses, is a political intervention into dominant development discourses that, by the end of the 1980s, were tied to a hegemonic environmental discourse. What ecofeminism allows in this context is a feminist intervention into changing development discourses as well as a location within which coalitions between southern and northern feminists can take place.

Let's begin by briefly outlining some of the more local as opposed to global problems with a U.S. ecofeminist discourse of racial and cultural diversity that privileges "international" difference instead of U.S. racial categories of difference. Within U.S. ecofeminist organizations, conferences, and writings, because the non–U.S. women who are used to construct "international" diversity within ecofeminism are often either of a privileged class in their home countries or are reductively constituted as "indigenous" women, "internationalism," as a U.S.-based discourse of cultural diversity, often elides important differences of class, caste, education, language, or culture that may be very pertinent within the home countries of non–U.S. women. Approaches that focus on questions of the specific interaction between U.S. racism and environmental problems are consistently displaced by the use of non–U.S. women to represent "diversity." And in conflating U.S. racism with U.S. neocolonialism, U.S. ecofeminists are impeded in offering a politically relevant, materially grounded analysis of the interaction between the two in the creation of environmental problems, whether they are seen as "local" or "global." In the U.S. context, critiques of ecofeminist essentialisms of race and gender that posit "indigenous" women as having a privileged standpoint in relation to environmentalism can be problematic. Nevertheless, as I argue in this essay, under specific historical conditions, ecofeminism has been an important *international* political location at the intersection of environmentalism and feminism, which has become a globalized space for political demands by women in many countries who might not otherwise have had a voice or an opportunity to create coalitions.

Critics such as Bina Agarwal, Cecile Jackson, and Brinda Rao have pointed out problems with the production of an "internationalist" ecofeminist movement. Primarily, their critique is focused on an essentialist discourse (which, contrary to their portrayal, is not singularly of ecofeminist origin) that sees symbolic "indigenous" women as the primary victims of the interaction between environmental problems and sexism as well as the inspirational sources of activist resistance to these problems.[1] It is important to note that although this is a discourse that can be found within ecofeminism—perhaps especially within a certain time period—it is neither solely ecofeminist nor unchallenged within ecofeminism.

These latter points are not widely understood. Instead, ecofeminism often serves as a straw-woman for a critique of a broader Western environmentalist discourse (in which, as we have seen, some ecofeminists are complicit through a complicated effort intended to construct an antiracist, anti-essentialist ecofeminism) about indigenous peoples as the "ultimate ecologists." However, given its status as a straw-woman in these debates, ecofeminism is clearly not the singular object of this critique; indeed, for Jackson and Agarwal, ecofeminism is a synecdochic figure for a discourse within development studies called "women, environment, and development," or WED. That is, these critics see a growing relationship between essentialist theories of women's stake in environmentalism (which they call "ecofeminism") and contemporary analyses within development studies of environmental problems and their solutions. For Jackson in particular, the main target is development discourses about women and the environment, even though most of her theoretical critique is directed against ecofeminism. This rhetorical move unfairly reduces ecofeminism entirely to an essentialist discourse and abstracts it from its historical and political context. For instance, Jackson writes: "How are ecofeminist ideas reflected in development literature and practice? . . . [I]t is taken as self-evident that harm to nature equals harm to women because of the pervasive perception that women are closer to nature. . . . The linkage of 'women' and environment is either simply assumed or asserted and used to prescribe actions to mobilise women for conservation."[2] This portrayal of ecofeminism as positing women as "closer" to nature is contradicted by many ecofeminist writers. For instance, Ynestra King, Karen Warren, Carolyn Merchant, and others posit women's relation to the environment as socially constructed and/or arising out of historical, materialist conditions; further, these writers see women's environmental mobilization as arising out of women's political agency rather than their essential similarity to nature. Rao, writing a little earlier than either Jackson or Agarwal, similarly locates problems of essentialism in what she sees, from her historical position, as an "emergent" set of studies (she does not immediately identify ecofeminist work in this category) concerned with the effects of a process she calls "capitalization of nature": i.e., "colonial and capitalist practices, and the so-called development schemes sponsored by international organizations like the World Bank."[3] These studies, Rao claims, whether they are dealing with women as victims of the capitalization of nature or as heroic environmental activists, "are based on almost identical conceptions of . . . the proximity of women to nature."[4] Note that Rao identifies specifically political conceptions of indigenous women-as-victims and therefore women-as-activists as the moment when essentialist notions are constructed. Whether the close relationship between women and nature is seen as biologically based or produced from women's material location in socially produced divisions of labor, Rao argues that these conceptions "perpetuate an essentialist construction of women and tribals based on nostalgic presuppositions of how they might have existed in some distant past."[5] Having sculpted this approach from "emergent" development discourses, Rao then identifies it with "ecofeminism" by using Shiva's work as representative of this position.[6] Here again, a reductionist move results in tagging all ecofeminism with the label "essentialist."

A much fairer rendition of the relationship between ecofeminism and development discourse—one that includes the internal contests within ecofeminism over essentialist notions of the relationship between women and nature—is given by Melissa Leach. Leach notes three strands within development discourse dealing with women and the environment: an ahistorical emphasis on women as the sole managers of natural resources; an ecofeminist argument about the negative consequences of Western conceptions of women and nature (conceptions that conflate and devalue them); and "feminist analyses of the effects of capitalist accumulation on women and the environment" that, unlike the first approach, are both materialist *and* historical.[7] What is an improvement over some of the other accounts is Leach's attention to the debate *within* ecofeminism over how to characterize the relationship between "women" and "nature." Unlike Jackson and Rao, she notes that there are "two rather different strands of ecofeminism which must be distinguished": one makes essentialist (sometimes biological) arguments, and the other analyzes various ideological constructs of women and nature as histor-ically and culturally located.[8] The second she finds potentially very useful for understanding the processes of "development." Agarwal also notes that there are both essentialist and anti-essentialist versions of ecofeminism, but she then goes on to insist that ecofeminism posits "'woman' as a unitary category and fails to differentiate among women by class, race, ethnicity, and so on. It thus ignores forms of domination other than gender which also impinge crit-ically on women's position."[9] Though she allows in a footnote that Ynestra King, in her later work, does not do this, she leaves out numerous ecofeminist arguments that argue for attention to racism, classism, and other forms of domination.

A common aspect, then, of these straw-woman accounts of ecofeminism in development discourse is that they rarely deal with the full diversity of ecofeminist positions and writers. Oddities of attribution and labeling thus occur frequently, and, interestingly for the discussion in this essay, they are often centered on Vandana Shiva's work. For instance, Leach relies heavily on Sherry Ortner's classic essay, "Is Female to Male as Nature Is to Culture?" as an example of an essentialist ecofeminist position. This is peculiar, since ecofem-inism postdates Ortner's 1974 essay by a good deal; while some ecofeminist theorists have used Ortner's arguments, many do not, including many with "essentialist" positions.[10] Leach also counts Shiva in both the "ecofeminist" strand of discourse within development studies and in the "feminist analyses of the effects of capitalist accumulation on women and the environment" strand, which she counterposes to "ecofeminism." Rao, as we have seen, uses Shiva's self-labeling as an ecofeminist to tag "emergent discourses" about the "capitalization of nature" as essentialist. As I argue throughout this essay, I think this difficulty in fixing a definition of essentialist ecofeminism, or of ecofeminism as a whole, or of Shiva's work in particular, lies in the fact that ecofeminism in development discourse is not so much an immutable set of theoretical positions as it is a political intervention that continually shifts its discourse in relation to its negotiation with dominant forces in development politics.

What none of these accounts captures is the various political positioning within development studies and international political structures allowed by the ambiguity of the "ecofeminist" position. I do not want to ignore or dispute the dangers of essentialist notions of women, indigenous peoples, cultures, or nature, which critics like those I've discussed above have analyzed so well. Rather, I wish to point to the positive potential for ecofeminism as a strategic discourse within a particular historical moment in international politics. The "discourse of indigeneity"—when coupled with claims about women's stake in environmentalism, which I have identified as a problematic element in U.S.

ecofeminism in the late 1980s and early 1990s—as an *international political discourse* rather than a *theoretical tool* opens up some possibilities. First, it creates a space within which southern women are authorized as experts. Second, the feminist and antiracist intentions of most ecofeminists exist in tension with their desire to positively revalue nature, women, and indigenous peoples. This contradiction produces opportunities not just to concur with but also debate those essentialist notions of women and nature that already may be circulating within masculinist development discourse, particularly if southern feminist environmentalists are included in coalitions. Finally, ecofeminism inserts feminist demands and analyses within a hegemonic discourse of globalizing environmentalisms at an important historical moment and to do so, it must at least momentarily posit a political collectivity called "women." This is a level of analysis not available in the criticisms of Agarwal, Jackson, and Rao. Leach, though not situated at this level, does note that "when policy-oriented discussions incorporate ecofeminist ideas they often mix [essentialist and anti-essentialist variants] uncritically,"[11] but she does not identify that mix as an opportunity for a strategic notion of ecofeminism that can be inserted into dominant political discourses and that contains the seeds of destabilization of its own (and the dominant discourse's) essentialism.

A less reductive story is told of the interweaving of ecofeminism and development discourse approaches by several books on women and development.[12] From this angle, we can see another origin story for ecofeminism: as an international movement rather than a U.S. movement. The texts and events that I examine in this essay map the intersections between feminism and development that produce an intermingling of "ecofeminism" and "women in development," which in turn produces, variously, the disciplinary, policy-oriented locations "women, environment, and development" or "gender, environment, and development." Discourses concerning the issues of women, development, and the environment were brought together in an uneven, contradictory, and contested process of negotiation over the production of knowledge, the distribution of resources, and the moral underpinnings of various policies and practices within an international political arena.

Sorting out U.S. ecofeminism's part in this complicated process is a difficult task, but it is clear to me that the kind of antiracist desire that produces the discourse of "indigenous ecofeminists" has its counterpart internationally in environmental feminist efforts to refigure "development" as "sustainable development," and then "sustainable livelihood," within a process in which the categories of "indigenous peoples" and "women" come to have a good deal of discursive, political, and moral weight.

To unravel this history, we need to start acronymically, with a tale of WID in conflict with GAD, WED becoming GED, UNEP organizing SWAGSD which influences the FLS, which begets WEDO which intervenes in UNCED; a story that interweaves international and internally contested movements of environmentalism and feminism with the machinery of UN bureaucracies, state structures, and multinational corporate interests.[13]

WID, or "women in development," is the name for a shift in development studies and policies in the early 1970s from women being invisible or appearing only as housewives and mothers, relegated to a privatized notion of reproduction, to seeing them as producers and economic actors, especially in the area of agriculture. This shift was brought on by the publication of Ester Boserup's 1970 book, *Women's Role in Economic Development*, which argued, influentially, that women were crucial to agricultural and commodity production in peasant Third World societies, and thus that development policies that failed to include them or to study their roles were doomed to failure. Boserup's goal was to use development efforts to increase gender equity in Third World societies, as part and parcel of the process of modernization.[14] Her approach, which did not challenge the foundational ideas of development itself (that modernization, Western-style, was a good and inevitable thing, and that Western experts should be the leaders in constructing development policy and Third World people the recipients), was crucial in producing new studies of Third World women's productive roles. However, the resulting policy programs did not often stress the need for gender equity. Rather, women were subjects for research which was aimed at creating more "efficient" and "effective" development policies, and their work was seen as an important resource for the success of development projects, even when those policies benefited men far more than women.[15] Indeed, as Leach points out, essentialist notions of nature and women, especially poor and rural Third World women, were common in WID discourse before "ecofeminism" became a player within development politics.[16] Nevertheless, WID was an important location for an internal contest between feminist notions of equality and empowerment and the desire of First World development agencies to craft policies that would successfully export Western products and practices to Third World countries. In practice, WID policies often provided poor women in the Third World with substantial opportunities compared to the previous male-oriented development paradigm.

Indeed, the WID paradigm was intimately intertwined with the growth of an international feminist movement, thus bringing international feminism into close contest and negotiation with Western multinational and state powers. The WID approach became initially institutionalized in the "development bureaucracy" in the North.[17] In 1975, the First UN Conference on Women and Development was held in Mexico City. Though the growing legitimacy of WID was not the only impetus behind this conference, it certainly was an important factor in convincing the UN's international policy makers that a conference on women was needed. By this time, in Sabine Haüsler's words, WID "was a more or less respected area of study; the number of publications on women and development topics has increased steadily ever since. Women and men sociologists and anthropologists, as well as a slowly increasing number of women development professionals in technical fields . . . from both North and South, moved into the field of development work."[18]

Coinciding with the Mexico City conference on women, southern women began to be heard more effectively, often as researchers themselves within the field of WID, at the same time as the approach became increasingly influential in development studies. During the Decade for Women, which was initiated by the 1975 conference, both northern and southern feminists began the process of constructing an international feminist movement, in which development policies—especially those that exploited the South by the North—were bones of contention for southern feminists, who accused northern feminists of ethnocentrism and of being tools of Western neocolonialism. In the 1980s, during a period of worsening conditions for women worldwide caused by the "debt crisis,"[19] southern feminists were organizing to influence the international political processes surrounding the UN apparatus that had grown up around women as political and economic subjects. In 1984, an organization called "Development Alternatives with Women for a New Era," or DAWN, was created. DAWN critiqued the WID approach for its acceptance of the "Western development model" and its failure to focus on the empowerment of women as a primary goal.[20] In Nairobi in 1985, during the Second UN Conference on Women, which culminated the Decade for Women, the parallel Non-Governmental Organization (NGO) Forum involved numerous lively discussions of the need for approaches such as DAWN's.

Many other southern analysts urged WID scholars to alter their approach from one in which women were simply "added on" to existing analyses to one in which gender relations, rather than women as essentialized objects of study, were the focus. This outlook, which aimed to contextualize cultural specificities through—and illuminate power inequities within—gendered relationships as manifested in household forms, marital customs, and gendered patterns of land ownership and use, was called GAD, or gender and development. Thus the shift from "women" to "gender" was meant to de-essentialize theoretical understandings of women's roles in their various societies and to reintroduce the feminist imperative of changing unequal gender relations along with promoting "development."

Concurrent with the rise of WID was a growing interest in environmental questions as part of development studies, which also included a focus on women—in part because of the new stress on women's management of natural resources through their productive roles, and in part because of an early link made between environmental problems and population growth. Haüsler notes, for example, that the "oil crisis" of the 1970s spurred development experts to look closely at the use of firewood by peasant Third World societies for energy. Since, within the WID paradigm, it was clear that women were the primary fuel gatherers and consumers of firewood, a link was quickly made between women, environmental degradation, southern population growth (a major obsession of the "Limits to Growth" report), and poverty. As Haüsler comments: " A powerful image emerged of poor people in the South, with too many children, using too much fuel; the poor were seen to have no choice but to destroy their own environment."[21] As a result, as Western consciousness of environmental questions such as resource use, energy production, desertification, and pollution increased during the 1970s, Third World rural women became scapegoats within development planning. The responsibility for population growth and environmental problems was thus placed on poor Third World women rather than on Western industrialized nations that consume most of the world's resources. These assumptions about Third World women's responsibility for dangerous levels of population growth and misguided environmental resource use, widespread in influential reports such as "Limits to Growth," have been a major target of attacks both by southern and northern feminists.

Several events served to embed the relation between environment, women, and development both within development politics and within international feminism as they interacted during the late 1970s to early 1990s. One could offer many examples of how the two streams of research, policy formation, and movement struggle focused on women and on the environment are not separate but rather interactive from the start. Important environmental paradigms are referenced by the "Limits to Growth" report (the problem of possible scarcity brought on by population growth coupled with resource use) and the Brundtland Commission's "Our Common Future" (which sketches the need for "sustainable development"). Both of these paradigms were challenged by feminist demands to reconceptualize women as having agency and needing power in environmental decisions. The feminist agenda became not only to make women's stake in these issues visible but also to promote women's economic and political empowerment. At the first major international conference on the environment, in 1972 at Stockholm, during the parallel NGO meeting, Sunderlal Bahuguna, a male Indian activist, presented Chipko women as exemplars of community-based, sustainable environmental practices. Because, as Haüsler writes, "women had emerged as the main actors in this movement it was concluded that rural women understood that it was in their own interests to protect the environment."[22] The Chipko movement was inserted into the international political context at a moment when the environment became a major agenda item, and in Bahuguna's version, Chipko represented a southern challenge to the notion that Third World women are problematic environmentally; instead, he presented them as natural environmentalists. It is this opening that was later seized by ecofeminists such as Shiva, who pushed the Chipko to represent women not just as natural environmentalists, but women as active, political agents with expert knowledge about the environment. Thus, "ecofeminism," in the late 1980s, entered the international context by attaching earlier feminist efforts to transform WID to a newer environmentalist paradigm, which, as I discuss shortly, was becoming at the time a hegemonic global formation. In this context, "ecofeminism" *means* this feminist intervention into environmentalism more than it represents a set of new, independent theoretical arguments.

While this "ecofeminist" position thus has political relevance and effectiveness within a political context at a particular time, it also runs the risk of dovetailing with older WID assumptions about women as "natural resources." There was concern that ecofeminist arguments (such as Shiva's) defining women as environmental managers and activists would be translated into development policies that required women to be the primary laborers in conservation schemes that may or may not have benefited them directly. Many progressive development scholars critiqued these arguments as essentialist, though such essentialism had existed as well in older WID discourse.[23] In a pattern similar to the move from WID to GAD, a focus on women, environment, and development, or WED (which could also be seen as the "ecofeminist" moment), has been recently been challenged by GED, or gender, environment, and development, a position that pays attention to the nuances of gender relations in households, property rights, labor relations, and kinship systems, all of which determine a differential relationship between women and their environments dependent on age, marital status, and many other factors. GED scholars argue that these nuances must be taken into account in policy planning, and that assumptions of women's natural tendency to protect the environment are deeply misguided. It is this debate with which our critics, Agarwal, Jackson, Leach, and Rao, are concerned. But rather than see the recurrence of essentialist moments in development discourses on women being part of an ongoing process of political struggle stimulated by feminist interventions, these scholars critique "ecofeminism" instead.

Nevertheless, it is clear that the insertion of feminist demands into development policy in the period when it became concerned with the environment (in other words, the "ecofeminist" moment) provided particular political and working links between Western feminists and southern feminists. In this political context, "the environment" served feminists as a medium for the connection of critiques of militarism, capitalism, and neocolonialism—similar to the way "militarism" functioned in the 1970s and 1980s. Positioning women as environmental activists was one moment in a dialectical process of negotiation between dominant interests in development policies and feminist efforts to insert women's concerns into an international arena. While it is important to critique the limits of such ecofeminist interventions, it is equally important to see the way in which these analytic linkages can operate as "two-way streets" between southern and northern feminist activists.[24] This is particularly the case when ecofeminist arguments, as Leach pointed out, contain a tension between essentialist and anti-essentialist analyses, giving an opening for debates around operative definitions of "women" and "nature" that open the door for more nuanced analyses.

The argument that women have a stake in environmentalist politics became an especially important strategic position within an international context of what I call "globalizing environmentalisms." I give a more specific example of such "ecofeminist" positioning when I examine WEDO. But first I want to make a few points about the status of environmentalism within global political discourses at the end of the cold war.

How do we explain the apparent convergence between the approaches of women, environment, and development discourse within development studies and certain strands of ecofeminist discourse? We explain it as another example of a feminist intervention into a masculinist environmentalist discourse, but in this case, one embedded in ongoing contests within a process I call the "globalization of environmentalism."

Without disputing the accuracy of those critiques (especially when it comes to the need to produce feminist scholarship as a basis for effective, empowering policies) that identify the essentialism of the "ecofeminist" discourses deployed in the women, environment, and development arena, I want to take a brief look at the wider context of this deployment. An examination of the context of "globalizing environmentalisms" during the end of the 1980s and beginning of the 1990s sheds light on the importance of a deployment of ecofeminist rhetoric for the construction of an international feminist movement.

Environmental problems, it has been pointed out frequently, do not honor fixed spatial areas, whether they be defined as national areas or spaces of private property. This characteristic of environmental problems has been, at different moments, the source of environmentalist claims for the need for a new global cooperation as well as a deep pessimism about the possibilities of solving environmental crises. The optimism of the global environmentalists has a negative side, however, and that is the use that can be made of the "universalizing" momentum of environmentalism by forces of technocratic, exploitative, neocolonialist, neocapitalist political economies. Southern environmentalists, like Guha, have thus critiqued the ways in which consciousness of these environmental problems are "global" in another sense, that is, tools for colonialist projects of northern exploitation of southern peoples and lands.[25]

In a perceptive article entitled "How Do We Know We Have Global Environmental Problems?" Peter Taylor and Frederick Buttell sketch the growing influence of the characterization of environmental problems as "global." They identify two ways of talking about environmental problems as global that they call the "technocratic" and the "moral." Though in some ways these discourses may seem opposed, since the "moral," or Green, discourse is putatively ranged against the scientistic, economically driven discourse of the "technocratic," Taylor and Buttel claim that there is a convergence between the two in that they propose a unitary human concern that avoids consideration of the varied material and political sources of environmental problems. They thus point out the possibility that the two discourses may operate together, imposing a dominant discourse which assumes the "sameness" of people, in order to achieve particularistic goals (i.e., those of multinational corporations, or Western elites). The idea that environmental problems are global and require global solutions, then, supports "either a moral response—everyone must change to avert catastrophe!—or a technocratic response—only a superintending agency able to analyze the system as a whole can direct the changes needed. There is no paradox here—moral and technocratic responses are alike in attempting to bypass the political terrain in which different groups experience problems differently and act accordingly."[26] Note that this objection to the "globalization of environmentalisms" parallels the ecofeminist objections to the philosophical "holism" of deep ecologists.

Taylor and Buttel locate the origin of the trend toward describing environmental problems as global as a specifically U.S. convergence between the scientific understanding of environmental problems and the environmentalist ("moral") response to them; indeed, at one point they cite the "long hot summer of 1988 in the United States" as one stimulus to the use of "global climate models."[27] But they also suggest another impetus to the globalization of environmentalism:

> The rise of global-change-led international environmentalism occurred during a significant shift of the political center of gravity of the industrial world toward neo-conservative regimes. Modern environmentalism has accommodated itself surprisingly readily to the free-market resurgence. While international environmental groups yet reserve the right to criticize the World Bank and related institutions about the environmental destruction that results *from particular projects* or *types of projects* . . . environmental groups have generally worked with the Bank/IMF in a surprisingly harmonious manner in implementing conservation/preservation policies and programs in the Third World. There is a key coincidence of interest. . . . the Bank and IMF gain legitimacy in the eyes of the citizens and political officialdoms

of the advanced (increasingly "green"-oriented) countries by helping to implement environmental and conservation policies, while the implied threat of Bank or IMF termination of bridging, adjustment and project loans is useful in securing developing country compliance with environmental initiatives. Given this relationship, most environmental organizations have been disinclined to take on the world debt crisis, the net South-North capital drain, and the international monetary order as being fundamental contributors to environmental degradation.[28]

Whatever the exceptions that could be taken to Taylor and Buttel's depiction of environmental organizations here, it is important to note that they are not positing a conspiracy but rather a convergence of particularly enunciated concerns that illuminate the contours of a specific conjuncture (one I am calling here "globalizing environmentalisms") within a hegemonic struggle for dominance. And we could enumerate the political struggles engaged in this conjuncture and identify them in multiple ways: between North and South, between class formations, between racial/ethnic groups, between genders, and so forth. Instead of this multiplicity, these struggles take place within internationalized versions of "environmentalism." What strikes me in Taylor and Buttel's historicizing of this hegemonic shift is something they don't mention: the relation of the "ending" of the cold war in the late 1980s to the appearance of global environmentalisms as a discursive tool within these political struggles. Like the discourse of democracy as a worldwide good, environmentalism similarly turns out to be a two-edged sword.

I will expand on the implications of this point briefly. The modern U.S. rhetoric promoting democracy within an internationalist political arena can be seen first as an antifascist and then as an anticommunist rhetoric, generated during World War II as a justification for U.S. military involvement and then refined after the war as a bulwark against the critique of capitalism. Democracy moves from being an oppositional labor-movement goal to being a statist rhetorical tool in the anticommunist repertoire, with the deployment of the rhetoric of "saving the world for democracy" used to further capitalist, imperialist projects. Like this discourse of global democracy, dominant environmentalist discourse makes similar claims about "universal human conditions," similarly reduced to individualist solutions (for instance, individual recycling, which, like voting as a solution to social inequality, is a form of liberal rather than radical discourse), similarly eviscerated of a critique of corporate agency in either the production of inequalities or environmental problems. Gaile McGregor comments: "The globalization process is inherent simply in the fact that we all speak the same 'language' of capitalism," and environmentalism is particularly deployed in this context.[29] Anna Tsing makes a similar point when she writes about environmentalism in the 1990s becoming a "leading edge of global civil society. In contrast to social ecologists working for social and economic equity, civil society environmentalists build their message on political *equality*. Since political equality in the 1990s is understood as a concomitant to the spread of markets, it becomes identified with the acceptance of social and economic inequity in the name of democracy."[30] Tsing goes on to describe the way environmentalism as a "strategic universalism" came to "seem a defining feature of the new transnational Europe." Since it could be argued that environmental activism on both sides of the Berlin Wall was partly responsible for its fall, "one could look to environmentalism for the coming together of the message of science, as universal principles, and the message of universal human rights in the necessity of democracy to preserve the world's health. Furthermore, environmentalism was advanced by the kinds of transnational and global organizations that could make ignorant and uncooperative states, with their entrenched local cultures of power, see the truth of these universals."[31]

Further evidence of the way in which globalizing environmentalism is being used to replace cold war rhetoric about global democracy is the way in which environmentalism is being grafted onto the "lost" project of militarism, which was centrally supported by "global democracy" discourse during the cold war period. For example, two recent *New York Times* articles specifically describe the way in which environmental problems have become the focus of new U.S. military endeavors. One article describes the growing U.S. military identification of environmental problems as the new threats to "national security" because of the way they result in "political instability."[32] The other describes the new use of spy satellites for identifying environmental problems, thus justifying the defense appropriation of money for these technologies.[33] The popularity of Robert Kaplan's construction of a connection between environmental disasters and the threatening "chaos" within African nations displays the way in which older cold war forms of U.S. racist and Western colonialist fears about the "barbarity" of the Third World are transformed into a concern with environmental disaster that unleashes new forms of "savagery."[34] Kaplan makes clear that, like the hegemonic discourse about democracy, the hegemonic discourse of global environmentalism can also be used to impose unjust conditions on the poor and the colonized, who are often represented in this discourse as part of the environmental threat. In some ways, the development of southern environmentalism is a strategy precisely to resist these uses of global environmentalism, in recognition that the environment is now an important terrain of transnational political struggle.

Besides and within southern environmentalism, there have been feminist analyses that have critiqued this kind of hegemonic discourse. Like the feminist discourses that renamed "anticommunism" "militarism," these newer and related oppositional discourses identify "global environmentalism" as another project of "patriarchal capitalist maldevelopment."[35] In both discursive moves, the oppositional accomplishment is to point out the sexism of the appeal to a generic mankind and humanity and to uncover, instead of universality, the connections between sexism, racism, imperialism, classism, and, in both discourses, environmental exploitation. One example of this kind of deconstruction is the feminist analysis of Joni Seager, where she identifies the patriarchal characteristics of governments, militaries, and corporations as one of the main factors in their continuing responsibility for environmental destruction.[36] Another example of a feminist challenge to hegemonic discourses about development, democracy, and the environment is WEDO, which we examine below. It is important to note the continuity in these projects between a feminist critique of militarism and a feminist critique of environmental problems; both oppositional discourses are aimed at deconstructing the universalism of hegemonic discourses of either the cold war or of global environmentalism, and at showing their interconnections. In the process of this deconstructing, these oppositional feminist and ecofeminist discourses often construct their own "strategic universalisms," particularly through claims to be representing "women" as a unity. As Tsing says: "What is global essentialism good for? It is good, it seems, for arguing with other global essentialisms."[37] I give here a more specific example of this process of constructing "strategic universalisms": the organization WEDO, which, in the middle of this transition from the cold war to globalizing environmentalisms, constructs a recognizably ecofeminist intervention with a universalist bent.

In some ways, WEDO is an ironic figure in this essay; I use it to illuminate ecofeminism despite its refusal of the term. Like many of the grassroots women's activists who reject the label ecofeminist, WEDO avoids the name (but for different reasons) even while arguing for women's environmental action in "ecofeminist" terms. WEDO makes close connections between feminism, women's movements, and environmental activism and has several prominent ecofeminists in leadership positions. As was true for WomanEarth Feminist Peace Institute, the cofounders of WEDO came from a background of feminist antimilitarism. Yet, unlike WomanEarth, WEDO embraces political action other than direct action, directed toward traditional kinds of institutional change, in this case primarily within the UN. In particular, WEDO shies away from any connection between feminist spirituality and environmental politics. Further, unlike WomanEarth's focus on U.S. racism, WEDO's attempt to create an organization that models diversity defined that diversity as primarily, though not exclusively, international.

WEDO was founded by two white U.S. women, Mim Kelber (activist and writer) and Bella Abzug (a former Democratic congresswoman from New York). Both had been founders of Women Strike for Peace and active in the second-wave women's movement.[38] After Bella Abzug left the U.S. Congress,

Kelber and Abzug formed the Women's USA Fund, Inc., which supported various subgroups, including one called the Women's Foreign Policy Council, "which was aimed specifically at getting equal participation of women in decision-making, related to women's rights, peace, and security."[39] Basically, the organization performed a networking function, printing a directory of U.S. women involved in foreign policy issues and international affairs in 1987. In 1988, Kelber and Abzug became more centered on the environment, moving easily from their previous concerns with nuclear disarmament to a more generalized concern with environmentalism. As Kelber puts it: "We began to realize in talking about the security of the earth, you're really talking about literally saving the earth from these man-made threats to the health of the planet." Inspired by the Gro Brundtland Report, "Our Common Future," and concerned about the lack of women in policymaking positions in the national and international institutions concerned with the global environment, Abzug and Kelber organized a briefing on the state of environmental problems worldwide for prominent women in the U.S. Congress and women's political organizations.

The organizing for this briefing brought Abzug and Kelber in touch with a number of women environmentalists from the Third World, as well as women working on environmental issues within UN agencies. Kelber met Joan Martin Brown, a staff member of the United Nations Environmental Program (UNEP) and head of a women's environmental network called WorldWIDE, which had published an international directory of women environmentalists and researchers. Kelber says that at this point (around 1989), "I began to realize the link between environmental protection, development issues, poverty, and property issues," and she and Abzug became committed to engaging these issues as interconnected. Brown told Kelber about a conference UNEP was organizing, in connection with the preparations for the upcoming United Nations Conference on Environment and Development, called The Global Assembly of Women and the Environment, in which women from all over the world were going to be brought together to illustrate environmental "success stories." In line with the image of women as resources for environmental programs, which was prominent in the WID development discourse at that time, the success stories were meant to demonstrate various alternative technologies and environmental practices developed by women in their communities, but not necessarily to generate influence on UNCED at the governmental, policymaking level. Brown suggested to Kelber that the Women's Foreign Policy Council might organize another conference at the same time, which would be more confrontational, more politically challenging, and more policy oriented; one that would deal, as Kelber put it, "with the larger issues."

Kelber and Abzug took up this project, which became the World Women's Congress for a Healthy Planet, held in Miami in 1991. In 1990, to prepare for this effort, Abzug and Keller brought together a fifty-member committee, called the International Policy Action Committee (IPAC), which decided on the name Women's Environment and Development Organization for the entity organizing the conference and chose ten co-chairs to direct WEDO. A serious effort was made to create international diversity in both IPAC and WEDO; the co-chairs were Bella Abzug (United States), Peggy Antrobus (Barbados), Thais Corral (Brazil), Marla Eugenia de Cotter (Costa Rica), Elin Enge (Norway), Farkhonda Hassan (Egypt), Wangari Maathai (Kenya), Chief Bisi Ogounleye (Nigeria), Vandana Shiva (India), and Marilyn Waring (New Zealand). These women were already active in development politics, promoting the perspectives of southern women (Peggy Antrobus, one of the founders of DAWN; Thais Corral, information officer for DAWN; and Chief Bisi Ogounleye, a vice president of the Forum of African Voluntary Development Organizations); environmental sustainability (Elin Enge, director of the Norwegian Forum for Environment and Development; and Wangari Maathai, founder of the Kenyan Green Belt Movement); and international feminism (Marilyn Waring, who had been the executive director of the Sisterhood Is Global Institute; and Farkhonda Hassan, chair of the executive committee of the Gender, Science, and Development Program at the Institutes for Advanced Study in Toronto). All, therefore, demonstrated long-term commitments to the intersections of the issues of development, environment, and feminism. Some were also connected with an explicit ecofeminist perspective (especially Shiva, through her inclusion in and authorship of ecofeminist publications; and Thais Corral, who is an editor of *Eco-Femina*, a radio program broadcast throughout Brazil and sponsored by UNIFEM).[40]

The World Women's Congress for a Healthy Planet brought together 1,500 women from eighty-three countries, with about one-third coming from developing nations.[41] An important feature of the conference was that it took the form of a tribunal: in front of five judges (Justices Desiree Bernard of Guyana, Elizabeth Evatt of Australia, Sujata V. Manahar of India, Effie Owuor of Kenya, and Margareta Wadstein of Sweden), "witnesses" gave testimony over three days. The topics of each day were "Towards Earth Charter '92: Developing a Code of Ethics with a Women's Dimension"; "Saving Natural Systems: Environment and Positive Development"; and "Science, Technology and Population." The witnesses, experts on their topics from a wide variety of countries, gave their testimony before the assembly of 1,500 women, many of whom were women brought as part of the "success stories" conference organized earlier by UNEP, and thus often part of grassroots efforts in their communities.[42] Other attendees at the conference were women who were part of environmentalist and feminist movements and organizations from around the world.

The format of a tribunal created a powerful experience. As each speaker gave evidence of the costs of development policies and environmental degradation for women in her country, the sense of injustice and outrage mounted. One reporter summarized some of the "horror stories":

> In a Malaysian village, where a Japanese consortium sold radioactive waste as "fertilizer" for home gardens and window-boxes, children are now dying of leukemia.
>
> From famine-stricken Ethiopia came news that the country, once sixty percent covered with forests, has only three percent left.
>
> From Tibet came an urgent plea for help in stopping the suspected Chinese dumping of nuclear waste into Tibet headwaters, "threatening the seven great rivers of Asia."
>
> From Bhopal to Chernobyl to the "Triangle of Death" in heavily polluted Eastern Europe, women report birth abnormalities and high levels of sickness among children.
>
> On a coral atoll in the Pacific, where, according to Marilyn Waring, "the French still insist that their nuclear testing has no effect on the food chain, women hang fresh fish like laundry on a line. They eat only those that flies land on; the others are discarded as too contaminated even for flies."[43]

The testimony during the conference covered many topics, including "Poverty, Maldevelopment, and the Misallocation of Resources"; "Earth's Refugees: The Causes of Uprootedness and Global Homelessness"; "Ethical Considerations of Nuclear Power and Weapons, and Other Threats to Public Health and the Environment"; "The Appropriation of Tribal Lands by the U.S. Government"; "Biotechnology and Biogenetics"; "Population Policies, Family Planning, and Sexual Politics."

In organizing the conference, as I have said before, WEDO wanted to generate a diversity defined by international, rather than U.S. racial difference. As a result, most of the U.S. women attending were white women,[44] and representation was distributed by world region rather than by racial difference within the United States. Though the conference opened with a "Traditional Call of Welcome" from "Indigenous Women of Florida," the only other featured speakers who were U.S. women of color were Faye Wattleton (president of Planned Parenthood at the time) and Winona LaDuke, Anishinaabeg feminist activist. LaDuke has expressed ambivalence about seeing herself as part of ecofeminist—or white U.S. feminist environmentalist—movements. After LaDuke's name in the initial program it says "Invited," as though marking her ambivalence.

Asked about the relative invisibility of U.S. women of color in the conference, Mim Kelber explained that WEDO had raised money specifically to bring women from the Third World, including in their number the women brought to the earlier UNEP conference. This combination of funding accounted for the high number of Third World women participants, one-third of the conference attendees. However, a specific effort to bring Asian American, Latino American, and African American women to the conference did not result in significant funding. Kelber says:

> We tried, but we couldn't get enough funding for it. . . . funders like to fund activities in other countries, but not so much in this country. And I think it's a constant issue for us in the international environment and development movement. They [funders] tend to count it [diversity] in North/South terms and I don't think there's enough emphasis on class differences. I think it's very important for us to recognize that we have a lot of poor people in this country and a lot of illiterate people and a lot of people in great need who are being exploited by the same global economic forces that are exploiting these other countries. . . . our Third World within our borders. . . . [W]e really ought to organize a reverse solidarity campaign, that women in other nations should be expressing solidarity with women in the U.S. who are under such severe attack right now by the right wing.

Though U.S. women of color were more or less invisible within the World Women's Congress for a Healthy Planet in terms of featured speakers or the overall number of participants, there was a caucus of Women of Color of North America at the conference. This caucus made efforts in particular ways analytically separate from some of the other problems discussed by the conference to raise the visibility of the effects of U.S. racism, and, more specifically "environmental racism." This caucus's statement was published in the conference proceedings along with those of other caucuses, which otherwise were organized by geographic region with two specific exceptions, Women of the South and the International Indigenous Women's Caucus.[45] In this way the geographic categories that WEDO used to organize the conference were disrupted by three interventions: one centered on U.S. racism, one on Western colonialism, and one challenging the first two by identifying indigeneity as an independent identity, occupied by members of the "Fourth World." Thus, U.S. racism and Western colonialism were tagged as structuring inequalities and promoting suffering along the lines of three notions of difference separate from geographic regional difference: "nonwhite," "underdeveloped," and "indigenous." The formation of these caucuses created a situation in which each process of identity formation within the conference—by gender, by nationality, by relation to colonialism, by race—was thereby destabilized.

Interestingly, the statement of the International Indigenous Women's Caucus is entirely devoted to promoting a spiritual relation to the earth that promotes the well-being of human and of nonhuman nature. "As indigenous people our lives are intertwined with the natural world," the statement begins, and adds: "Today we face the destruction of the human spirit and the consequent destruction of the natural world. . . . The true challenge of human beings is to place our full attention upon ways in which we can live upon Mother Earth in a manner consistent with natural law and in peace, harmony, and balance with all living things."[46]

Clearly, the politics of the World Women's Congress for a Healthy Planet intertwined feminist and environmentalist positions with a number of radical analyses. This could be called an "ecofeminist" politics, similar to the ecofeminism developed by such writers as Karen Warren, Val Plumwood, and Ynestra King, but one grounded in an exploration of women's daily problems and material constraints and presented within a framework of international diversity. Two things created coherence for the bringing together of such different issues as women's struggles with nuclear contamination, the effects of imposition of debt dependence on poor nations, and coercive population policies: an analysis of the interconnection of multinational capitalism, sexism, colonialism, racism, and environmental exploitation and a rhetoric locating women as the primary victims of these forces as well as the most effective political agents against them.

WEDO's published materials construct a unity for women based on their exclusion from male-dominated policymaking institutions worldwide, as well as on their social roles as caretakers. At the beginning of the creation of WEDO, in 1989, the organizers published a statement entitled "A Declaration of Interdependence."

Referencing the original Declaration of Independence, in a protest tradition of reworking that document that goes back to the 1848 "Declaration of Sentiments" of the nineteenth-century women's movement, the Declaration displays a perspective recognizably ecofeminist. Using one of the favorite metaphors of antimilitarist feminism and ecofeminism—the web—the Declaration argues that sexism and environmental degradation are ideologically and materially linked: "It is our belief that man's dominion over nature parallels the subjugation of women in many societies, denying them sovereignty over their lives and bodies. Until all societies truly value women and the environment, their joint degradation will continue." Further, the Declaration insists that joining feminist and environmentalist perspectives demands attention to diversity along several axes, as well as to militarism, poverty, and political equality.

WEDO's closeness to an ecofeminist perspective also can be seen in the language of "Women's Action Agenda 21," which was the culmination of the

World Women's Congress for a Healthy Planet in November 1991. The document was used as a manifesto for a feminist intervention into the process of UNCED, held in Rio in June 1992. Its title referenced *Agenda 21*, which was the document to be produced by the governmental bodies in Rio. Before and during the period that women met at the Congress for a Healthy Planet in Miami, other preparatory meetings to shape *Agenda 21* were taking place around the world under UN auspices. Bella Abzug had attended many of these meetings and, alarmed at the lack of participation or power of women in these meetings, had in each case brought those women present together in a "women's caucus," designed to articulate the collective needs of women in relation to the environment and development issues being discussed. This strategy, or "methodology," as Abzug called it, helped to provide WEDO with the personal contacts and sense of the issues that became the basis for the Congress for a Healthy Planet.[47]

The "Women's Action Agenda 21" contained a list of specific demands aimed at the governments participating in UNCED, organized around the topics "Democratic Rights, Diversity and Solidarity"; "Code of Environmental Ethics and Accountability"; "Women, Militarism and the Environment"; "Foreign Debt and Trade"; "Women, Poverty, Land Rights, Food Security and Credit"; "Women's Rights, Population Policies and Health"; "Biodiversity and Biotechnology"; "Nuclear Power and Alternative Energy"; "Science and Technology Transfer"; "Women's Consumer Power"; and "Information and Education."[48] It also issued deliberate challenges calculated to address the lack of women's power within the UN itself as well as the UNCED. The document required that a "permanent gender-balanced UN Commission on Environment and Development" be created; that the imbalance of gender ratios in the UN staff, especially in agencies like UNEP, be redressed; that donor countries increase their funding of UNIFEM (a UN fund for providing resources for and research on women's issues); and that member nations send to UNCED gender-balanced delegations, which would also include representatives of indigenous peoples and grassroots organizations. The creation of a *women's* "Agenda 21" at the Congress required debate and agreement among the 1,500 women present (a process managed through workshops on various issues, in which participants agreed on language to be presented to the larger group). Thus, it is a statement carrying a lot of weight, representing the strong coalitions built among women from very different political and cultural locations, across national borders.

Consolidated by a similar process of debate and agreement among diverse women from around the world at Planeta Femea, or the Women's Tent at the Global Forum, the NGO alternative gathering at the UNCED meeting in Rio in 1992 reworked the document for presentation to the official governmental bodies.[49] In this context, the "Women's Action Agenda 21" was used to lobby for "women's issues" to be included in the formal *Agenda 21*. Indeed, this lobbying effort succeeded in getting specific mention of women's issues in terms of the political perspective of "Planeta Femea" in thirty-three of the forty-plus chapters of *Agenda 21*, not counting the inclusion of a chapter specifically addressing the importance of considering women as agents of environmental change as well as the relation between sexism and environmental degradation. This chapter, entitled "Global Action for Women Towards Sustainable and Equitable Development," incorporated the political perspective fostered by WEDO organizers into the heart of the formal government agreements and represented a significant feminist intervention into development politics and the sphere of globalizing environmentalism.[50]

The Preamble of the "Women's Action Agenda 21" couches the diverse issues discussed in the Congress for a Healthy Planet in language that constructs women as activists on behalf of the environment through their commitment to justice, equality, and nurturing. With a recognizably ecofeminist voice, the Preamble argues for interconnections between various political struggles, stating that "a healthy and sustainable environment is contingent upon world peace, respect for human rights, participatory democracy, the self-determination of peoples, respect for indigenous peoples and their lands, cultures, and traditions, and the protection of all species." These things are connected for the writers of the Preamble, because "as long as Nature and women are abused by a so-called 'free-market' ideology and wrong concepts of 'economic growth,' there can be no environmental security."[51] The correspondence is exact between WEDO's arguments here, representing all of the women at the conference, and ecofeminist explorations of the consequences for the environment and for women of how, in Western ideology, women have been equated with nature and both have been devalued.

Despite the similarity between the analysis and rhetoric of WEDO and ecofeminist writing of the time, WEDO organizers gently shied away from the label "ecofeminism." Exploring this aspect through interviews with Mim Kelber and Bella Abzug, I found that reluctance embedded in the notion of ecofeminism as a countercultural politics, or a politics based on feminist spirituality, a politics less concerned with institutional politics and more with philosophical argument or direct action. In addition, Kelber and Abzug both expressed that ecofeminism, as they understood it, was about connections between women and nature and was too much a single-issue movement, unable to address the structural processes that produced women's inequality and environmental degradation. However, Kelber and Abzug did not display a thorough knowledge of the complexity of some ecofeminist arguments. For instance, when I pointed out to Kelber that, while WEDO did not use the term "ecofeminist" to describe itself, still many people perceived the "World Women's Congress for a Healthy Planet" as an ecofeminist conference, she replied:

> As you said, I'm not quite sure what the definition of ecofeminism is. The women's movement, including the feminist movement, has a lot of diversity and a lot of different interpretations. There's the whole spiritual group and the cultural group and then there are those of us who come into it via the political project. So we tend to work in a different way. Because there are two approaches, that is, that you create a counterculture, and that's what you do. And then there are others of us who feel that you have to try to make the system work for all of us. It's a reformist approach. I appreciate a lot of what women ecologists are doing and writing [about], talking about alternate ways of living and alternate systems and so on, but unless the global economy self-destructs, and that very well might happen, we have to deal with how the global economy is operating. . . . But I would say that Bella's and my approach has been that you try to work with what's real and change it. But we also understand that women have all different kinds of feelings, different interpretations. And whatever they want to do, it's okay. They can all feed into this so that we can have a common meeting ground on many issues. Our work styles may be different, but I think we all have this common goal of a healthy planet. And the core of it, and I guess Bella and I are very strong on this, is equality for women in decision-making.

Kelber's emphasis on a strategy of reform within existing political and economic institutions is what led Greta Gaard to remark, after her attendance at the Congress for a Healthy Planet, that WEDO was a "liberal ecofeminist" organization (which, to her, "sounds like an oxymoron").[52] Yet what interests me here is that Kelber, while constructing ecofeminism as a countercultural politics *not* interested in "reform," nevertheless sees the rhetoric WEDO employs as creating an umbrella under which many different kinds of women and feminists can create a coalition. The idea that "women all have the common goal of a healthy planet" reverberates within much of WEDO's literature, which relies upon the notion that women's equal participation in political decisions about policy will produce more environmentally sound practices. As Abzug notes: "We do regard . . . a clean environment, a healthy society, the preservation of the earth as being a very fundamental thing for any society, and we feel that women particularly have a very essential role to play."[53]

When asked about the manner in which this connection is made between women's empowerment and environmentalism, Abzug and Kelber articulated two notions: first, that women's particular social and material labor means that environmental issues are important to them (through their roles as mothers, health workers, and food producers); and second, that women have a different sense of connection to nature than men (ironically, the belief that

they articulate as belonging to "single-issue" ecofeminism). Often these notions come through in the course of their insistence that giving women an
equal say in governmental policy making would make a difference. On this
point, Kelber remarks:

> We keep saying we're not romanticizing women and demonizing men, but I
> think growing up female and growing up male is just different. It is different. I'm
> not talking about ability or brain size, or the right side of the brain or the left side
> of the brain. It's different, it's just all absolutely different experience. Some
> women may be able to totally separate themselves from the whole weight of tra-
> dition and social roles and so on, but for the majority of women in the world cer-
> tainly, they still bear the weight of the past.

And when asked about her reaction to some feminist criticism of the connection ecofeminists have made between women and nature, for instance in
the Women's Pentagon Actions, Abzug responds:

> Some people think that the emphasis on ecofeminism, by ecofeminists, on the
> natural bond between women and the earth is unacceptable to them. . . . I am ba-
> sically not an ecofeminist but the point is, I see, there is something that springs
> from the earth, there is a life, there is a nurturing, there is a symbiotic sense of
> preserving, and I've often said that as long as discrimination and degradation
> continue, [as long as] we continue degrading the earth, that we are at the same
> time creating a discrimination against women. So I think there is a symbiotic
> relationship.

Whatever the source of the connection made between women's issues and
environmentalism, WEDO organizers clearly feel that an appeal to women as
a collectivity, to their similarities despite their differences, is an effective organizing practice. Nevertheless, the politics underlying that appeal is one that
privileges the southern critique of Western versions of development, as well as
an interconnection between radical environmental, feminist, antiracist, and
anticapitalist analyses. For example, when asked why she thought of environmental issues as women's issues, Abzug replied: "I always think every issue is a
women's issue. I come from that school of thought . . . in fact, [when we had]
a congress which we called a World Congress for a Healthy Planet . . . we put
together not only our views on earth, air, soil and water, but our views of the
total environment, the environment of health, of human rights, of equal
rights, of political rights, of economic justice." And again, in explaining why
she does not want to restrict herself to being defined as all "ecofeminist," she
says: "I am not *just* an ecofeminist. Although we use some language which.
. . . brings us closer to that posture than most people . . . if you read the Preamble of our 'Women's Action Agenda 21,' we do think there is a bond be-

tween the earth and women. But we go much further than [ecofeminism], a much larger definition. . . . we are trying to include all kinds of people in our platforms of action and in our activities."

The rhetoric of WEDO thus moves between what might be called an essentialist ecofeminism, calling upon women in their roles as mothers and healers to take on environmentalist causes, and what might be called an anti-essentialist ecofeminism, paying attention to difference within a framework of analyzing the operations of political, economic, and social power. A poster that WEDO used in the early 1990s demonstrates some of the tensions within its political rhetoric. Under a stunning image of the earth seen from space are the words "It's Time for Women to Mother Earth." The text of the poster says: "With every day that passes, a little more of our world dies at the hands of pollution and neglect. But, *as women,* we can help do something about it" [my emphasis]. The text goes on to mention examples of women environmentalists active across the globe, such as Linda Wallace Campbell, active in the African American struggle against toxic waste in Alabama; Wangari Maathai, of the Kenyan Green Belt Movement; and Janet Gibson, who worked against the destruction of a barrier reef in Belize. The emphasis in the rest of the poster's text is on the need to bring women into the policymaking institutions that make decisions on the environment, rather than to take environmental action themselves. The assumption is that women will make more environmentalist decisions. And the assertion that "It's Time for Women to Mother Earth," while counting on women's maternalism, moves women from a symbolic, passive identity with Mother Earth to a position as active, political agents.

Notions of WEDO as a maternalist version of ecofeminism that sees women only in essentialist ways are disturbed by watching WEDO's "methodology" in action. The strategy of forming Women's Caucuses within the UN preparatory meetings assumed that women have something in common politically. But the emphasis within the caucuses on women's unity was most often constructed on the idea of women's exclusion from decision making and power, rather than on essentialist notions of maternalism. The practice of the caucuses insured that many women heard about other women's political struggles in an atmosphere of coalition-forming and respect. The chair of the caucus for each meeting was rotated, giving each region a chance to chair a meeting. During the caucus, agreement was sought on language that teams from the caucus (formed mostly by members of NGOs) would lobby to be included in the government documents agreed upon at the UN meetings. WEDO provided members of the caucus with the government document previously produced, and participants in thc caucus would try to agree upon language to amend these documents in directions favoring women and other unrepresented peoples as well as other political positions. If the Women's Caucus's lobbying efforts were successful, this new language would become part of the policy document agreed upon by all the member nations represented at the UN meetings. This "methodology" was so successful at Rio that WEDO continued it beyond the 1992 UNCED meeting that prompted its initial creation. WEDO was responsible for organizing Women's Caucuses before and during UN conferences on population in Cairo, 1993; and on social development in Copenhagen, 1994. Finally, WEDO also organized a caucus to provide a similar space for NGO coalitions for the Fourth World Women's Conference in Beijing, 1995; but since this conference was about women, it didn't make sense to call it a "Women's Caucus." Instead, it was called a "Linkage Caucus," a name that more accurately described, to my mind, what happened in all the preceding Women's Caucuses: the linkage of *issues* rather than *women*.

In these ways, WEDO's organizational format (not a binary construction of difference but choosing multiply located subjects for their involvement in the issues WEDO has identified as important) destabilizes its essentialist rhetoric about "Now It's Time for Women to Mother Earth." Despite the admirable construction of a politics of connection between important vectors of exploitation and injustice, and its success in influencing UN processes, WEDO's choice of the UN as its main focus of political action has obvious limitations. While an important international political arena, the UN itself has little or no enforcement capabilities to ensure that agreements made by governments during various global summits will be carried out. Though struggles over particular language in official transnational UN agreements are fierce and involve serious political issues, they end up as textual referents, often with their radical force significantly compromised by pragmatic realities of successful lobbying, rather than concrete political or economic practices.

Well aware of these limitations, though still committed to the practice of lobbying within the UN, WEDO has concentrated its energies after the Beijing conference on mobilizing women in their own countries to insist that their governments comply with the international agreements on women, environment, development, and population. One of the tools to further local organizing that WEDO developed early in its existence was the "Women for a Healthy Planet Community Report Card," which provides a framework and resources for community investigation and for publicizing local environmental and health problems. The "Report Card" contains guidelines on how to organize community organizations to create an "action agenda" around the suggested areas of "Natural Environment," "Political Systems," "Social Priorities," and "Human Development."[54]

Another aspect to assessing the effects of international strategies appears when attention is paid to the local and regional effects of the organizing initiated by international environmental and feminist groups. An interesting example of this process is reported by María Fernanda Espinosa, who as part of her work with the Indigenous Organization of the Ecuadorian Amazon (CONFENIAE) was asked to organize a Regional Workshop of Indigenous Women in preparation for the World Women's Conference in Beijing.[55] Noting that the "modified version of *Agenda 21*, considering specific recommendations for women, was an attempt to reconcile the sustainable development plan of action with the role and claims of women . . . to serve as a general framework for the deliberations on women and environment," Espinosa points out that "instead of building conceptual, political and operational connections between women, environment and development, the document has [only] superficial addendums." Indeed, Espinosa says, it could be called "a gender addendum not a gender agenda."[56]

Whereas at the international level, women's issues were superficially attended to, at the midlevel of the continental meeting called Encounter of Indigenous Women of the Americas, Espinosa found that women's issues were subsumed to the "struggle of indigenous peoples, about indigenous territorial and cultural rights, self-determination and bilingual education." While this articulation of indigenous politics would seem to include indigenous women's issues, in fact, according to Espinosa, the statements made by the participants at the Encounter revealed "the predominance of the ethnic emphasis over gender and environmental concerns as well as an implicit critique [of] western feminisms."[57]

Opposed to the international, continental, and national levels of organizing, Espinosa offers the story of the small, localized gathering of indigenous Amazonian women she was asked to organize by CONFENIAE. This effort was very new; CONFENIAE, in its thirty years of existence, had only paid attention to women as a special group when it was prompted to by the organizing efforts for Beijing. Her experience with the Regional Workshop of Indigenous Women ran counter to these tendencies to subsume women's self-perception of their needs and issues. Espinosa arrived at the gathering of forty-five Amazonian indigenous women, who were "local leaders with scarce political experience and often very little formal education,"[58] armed with paperwork to help the women construct their input into the continental group and therefore to the Beijing conference, only to have the women tell her they weren't interested in the UN preparatory process. They had never before been able to get together with others like themselves without elites (even their own elites) present, and they wanted to use the time to talk about how to help each other, to share information, and to strategize about local issues.[59] The document they produced to carry up to the international level was "based on personal testimonies and has a narrative form." Espinosa argues that

> This initial experience, in spite of coming more from external initiatives than from self-generated political needs, is encouraging indigenous women to establish an intercultural and intergenerational communication; to reflect about their needs and struggles; and think about the skills and alternatives they have to develop in order to face the changes and aggressive demands of the post-industrial world. Furthermore, this process may also contribute to the democratization of indigenous organizations themselves.[60]

Espinosa concludes from this experience,

> Looking to the three conferences [Beijing, the continental Encounter of Indigenous Women, and the local Regional Workshop], there is a disjuncture between the globalized discourse about indigenous women generated in Beijing and local initiatives. . . . What Beijing objectively did was to open spaces and opportunities for dialogue and communication at different scales. The incorporation of indigenous women's perspectives and political leadership in global agendas can be seen as one of the effects of the internationalization of the ethnic and gender debates.[61]

In his essay on the globalization of grassroots politics, Michael Peter Smith points out that in social theory, the local is usually equated with "stasis" and "personal identity," while the global is characterized as the "site of dynamic change, the decentering of meaning, and the fragmentation and homogenization of culture—that is, the *space* of global capitalism."[62] In contrast, he argues that a transnational grassroots politics has appeared, which confuses these older notions of separate local and global spaces, and which operates within particular transnational arenas. One example he offers is a hearing of "Bay Area migrant women" to give testimony to the UN Summit on Human Rights held in Vienna in 1993. Like the conference organized by WEDO in Miami and other gatherings like it, these activities create new transnational political subjects, brought together as women, or as members of other politically constructed subjectivities. They also create new opportunities for dialogue and coalitions.

These political collectivities very well may be constructed by essentialist discourses, but they are also collectivities built on hard-won unity across radical differences. And they may serve the less powerful groups as well as the powerful within the new collectivity. For instance, as I have argued in the case of WEDO, "ecofeminist" discourses about women's nurturing relation to nature intervene within hegemonic processes in a context of globalizing environmentalisms, and, through an organizational structure that emphasizes equal participation among very differently located political actors, serve to destabilize the essentialism of the rhetoric and produce valuable political effects. Just the construction of these arenas creates new opportunities for the less powerful to gain political leverage. Jane Jacob, in an essay critiquing essentialist Western notions about aboriginal women's relationship to environmental activism, makes a similar point: "In particular, there are specific problems arising from the essentialized notions of Aboriginality and woman that underpin radical environmentalisms and feminisms. Yet to read these alliances only in terms of the reiteration of a politics of Western, masculinist supremacy neglects the positive engagement indigenous women may make with such 'sympathizers' in their efforts to verify and amplify their struggles for land rights."[63] The ecofeminist intervention into UN processes creates a network, a space for debate, a mechanism not just for the intervention of feminism, environmentalism, and anticolonial scholarship into policymaking; but also for strategic coalitions to take place among disempowered people and between privileged and underprivileged people in one political collectivity. The practices and rhetoric of WEDO do not deal sufficiently with questions of the relationship of U.S. racism to environmental problems and to sexism. We still need organizational frameworks that can deal with this intersection. And we need more ways to intervene in globalizing environmentalisms besides the UN. But we also need to keep in mind that, as in the contribution of the international feminist antimilitarist movement to the end of the cold war, we may need to tolerate "essentialist" rhetoric that calls women from different locations to act together against power.

Notes

1. Brinda Rao, "Dominant Constructions of Women and Nature in Social Science Literature," *Capitalism, Nature, Socialism,* Pamphlet 2 (New York: Guilford Publications, 1991); Bina Agarwal, "The Gender and Environment Debate: Lessons from India," *Feminist Studies* 18 (1) (spring 1992): 119–58; Cecile Jackson, "Women/Nature or Gender/History? A Critique of Ecofeminist 'Development'," *Journal of Peasant Studies* 20 (3) (April 1993): 389–419.

2. Jackson, "Women/Nature," 399.

3. Rao, "Dominant Constructions," 12.

4. Rao, "Dominant Constructions," 17.

5. Rao, "Dominant Constructions," 18.

6. Rao, "Dominant Constructions," 17, n. II.

7. Melissa Leach, *Rainforest Relations: Gender and Resource Use among the Mende of Gola, Sierra Leone* (Washington, D.C.: Smithsonian Institution Press, 1994), 23.

8. Leach, *Rainforest Relations,* 30.

9. Agarwal, "The Gender and Environment Debate," 122.

10. Ortner's essay can be found in *Women, Culture, and Society*, ed. M. Z. Rosaldo and L. Lamphere (Palo Alto, Calif.: Stanford University Press, 1974), 7–88.

11. Leach, *Rainforest Relations*, 30.

12. I take as examples here, in order of their publication: Irene Dankelman and Joan Davidson, *Women and Environment in the Third World* (London: Earthscan Publications, 1988); Heleen van den Hombergh, *Gender, Environment and Development: A Guide to the Literature* (Utrecht: International Books, 1993); Maria Mies and Vandana Shiva, *Ecofeminism* (London: Zed, 1993); Rosi Braidotti, Ewa Charkiewicz, Sabine Haüsler, and Saski Wieringa, *Women, the Environment and Sustainable Development: Towards a Theoretical Synthesis* (London and Atlantic Highlands, New Jersey: Zed, 1994); and Vandana Shiva, ed., *Close to Home: Women Reconnect Ecology, Health and Development Worldwide* (Philadelphia: New Society Publishers, 1994).

13. The history that follows is deeply indebted to Braidotti et al., *Women, the Environment, and Sustainable Development*; and van den Hombergh, *Gender, Environment and Development*.

14. Braidotti et al., *Women, the Environment, and Sustainable Development*, 78. The chapter I reference in this section was written by Sabine Haüsler.

15. For a description of one effort to harness women's work to produce a successful development project benefiting men, see Richard Schroeder, "Shady Practice: Gender and the Political Ecology of Resource Stabilization in Gambian Garden/Orchards," *Economic Geography* 69 (4) (1993): 349–65.

16. Leach, *Rainforest Relations*, 25.

17. Sabine Haüsler, "Women, the Environment, and Sustainable Development: Emergence of the Theme and Different Views," in *Women, the Environment, and Sustainable Development*, ed. Braidotti et al., 80.

18. Haüsler, "Women, the Environment, and Sustainable Development," in *Women, the Environment, and Sustainable Development*, ed. Braidotti et al., 80.

19. See van den Hombergh's discussion in *Gender, Environment, and Development*, 58–60.

20. Haüsler, "Women, the Environment, and Sustainable Development," in *Women, the Environment, and Sustainable Development*, ed. Braidotti et al., 81.

21. Haüsler, "Women, the Environment, and Sustainable Development," 84.

22. Haüsler, "Women, the Environment, and Sustainable Development," 85.

23. Schroeder, "Shady Practice."

24. See Anna Tsing in "Environmentalisms: Transitions as Translations," in *Transitions, Translations, Environments: International Feminism in Contemporary Politics*, ed. Joan Scott, Cora Kaplan, and Debra Keates (New York: Routledge, 1997).

25. Ramachanda Guha, "Radical Environmentalism and Preservation of Wilderness: A Third World Critique," *Environmental Ethics* 11 (1) (1989): 71–83.

26. Peter J. Taylor and Frederick H. Buttel, "How Do We Know We Have Global Environmental Problems: Science and the Globalization of Environmental Discourse," *Geoforum* 23 (3) (1992): 405–416, 48.

27. Taylor and Buttel, "How Do We Know," 409.

28. Taylor and Buttel, "How Do We Know," 411–12.

29. Gaile McGregor, "Reconstructing Environment: A Cross-cultural Perspective," *Canadian Review of Sociology and Anthropology* 31 (3) (1994): 269.

30. Anna Tsing, "Environmentalism," 11.

31. Tsing, "Environmentalism," 12–13.

32. Steven Greenhouse, "The Greening of American Diplomacy," *New York Times* 9 October 1995, 4 (A).

33. "American Spy Satellites Will Turn More Attention to Nature with End of Cold War," *New York Times* 27 November 1995, A1, 6.

34. Robert D. Kaplan, "The Coming Anarchy," *Atlantic Monthly* 273 (February 1994): 44–46.

35. Shiva's term; see Vandana Shiva, *Staying Alive* (London: Zed, 1988).

36. Joni Seager, *Earth Follies: Coming to Feminist Terms with the Global Environmental Crisis* (New York: Routledge, 1993).

37. Tsing, "Environmentalism," 17.

38. My information about WEDO comes from Mim Kelber, "The Women's Environment and Development Organization," *Environment* 36 (8) (October 1994): 43–45; Carolyn Merchant, "Partnership Ethics: Earthcare for a New Millennium," in *Earthcare: Women and the Environment* (New York: Routledge, 1996), 209–24; various WEDO documents; interviews with Mim Kelber (March 1995) and Bella Abzug (May 1995); and participant observation of WEDO's activities in New York City at the March 1995 UN Preparatory Meetings for the Fourth World Women's Conference to be held in Beijing. I am deeply grateful to Mim Kelber and Bella Abzug for their time and frankness, and the WEDO staff for their assistance.

39. Mim Kelber interview, March 1995. All quotes attributed to Kelber in what follows are from this interview unless otherwise specified.

40. "Meet the Women Who Steer WEDO's Course, Our Co-chairs," in WEDO's newsletter, *News and Views* 6 (2) (September 1993): 2–4.

41. Kelber, "The Women's Environment and Development Organization," 43.

42. "Findings of the Tribunal," in *Official Report: World Women's Congress for a Healthy Planet* (New York: WEDO, 1991), 2.

43. Michele Landsberg, "Overview," in *Official Report*, 2–3.

44. See the special issue on the World Women's Congress for a Healthy Planet, *The Ecofeminism Newsletter* 3 (1) (winter 1992).

45. "Regional Caucus Reports," in *Official Report*, 26–35 (with inserted unnumbered page).

46. "International Indigenous Women's Caucus," in *Official Report*, 30.

47. Mim Kelber interview, March 1995; Bella Abzug interview, May 1995.

48. "Women's Action Agenda 21," in *Official Report*, 17–23.

49. Carolyn Merchant, "Partnership Ethics: Earthcare for a New Millennium," *Earthcare*; Rosiska Darcy de Oliveira and Thais Corral, *Terra Femina* (Rio de Janeiro: A joint publication of Institute for Cultural Action (IDAC) and The Network in Defense of Human Species (REDEH), 1992).

50. *Agenda 21: An Easy Reference to the Specific Recommendations on Women* (New York: UNIFEM, undated).

51. Preamble of "Women's Action Agenda 21" in *Official Report*, 16.

52. Greta Gaard, personal communication, letter dated March 12, 1996.

53. Bella Abzug, interview, May 1995. All quotes from Abzug are from this inter-view unless otherwise specified.

54. From "Women for a Healthy Planet Community Report Card," available from WEDO, 845 Third Avenue, 15th Floor, New York, NY, 10022.

55. María Fernanda Espinosa, "Indigenous Women on Stage: New Agendas and Political Processes among Indigenous Women in the Ecuadorian Amazon," paper presented at the Feminist Generations Conference, Bowling Green State University, February 1996.

56. Espinosa, "Indigenous Women," 3.

57. Espinosa, "Indigenous Women," 4–6.

58. Espinosa, "Indigenous Women," 10.

59. I am paraphrasing the story Espinosa told as part of her presentation to the Feminist Generations conference. This story is not in the text of her paper.

60. Espinosa, "Indigenous Women," 9.

61. Espinosa, "Indigenous Women," 10.

62. Michael Peter Smith, "Can You Imagine? Transnational Migration and the Globalization of Grassroots Politics," *Social Text* 39 (summer 1994): 15–33.

63. Jane Jacobs, "Earth Honoring: Western Desires and Indigenous Knowledges," in *Writing Women and Space: Colonial and Postcolonial Geographies*, ed. Alison Blunt and Gilliam Rose (New York: Guilford, 1994), 169–96.

7

Environmental Protection as Religious Action: The Case of Taiwanese Buddhist Women

Wan-Li Ho

Introduction

Wan-Li Ho, a scholar of religion, examines religious Buddhist women's environmental activism in Taiwan. She explores the women's environmental protection movement *(huan-bao)* showing how religious women have served as a catalyst for both environmental concerns and gender transformations. The result of her research provides another ecofeminist vision based on family ties and interreligious cooperation. Again we see that ecofeminism needs to be shaped by specific contexts, religious influences, and cultural patterns.

T HIS ESSAY EXAMINES RELIGIOUS (mostly Buddhist) women's activism in Taiwan and their environmental concerns, with a special focus on the 1990s.
It explores the women's *huan-bao* (environmental protection movement) by
way of case studies rather than social theory. I attempt to show how religious
women have served as a catalyst for both environmental concerns and gender
transformations and how a non-Western perspective within a specific national, historical, and multicultural context has opened another vision for
ecofeminism. The two organizations studied here are both from the Buddhist
tradition and were both founded by and led by women: the Buddhist Compassion Relief Tzu-Chi Foundation (Tzu-Chi) and the Life Conservationist
Association (LCA).

These case studies also tie in with recent industrial advances in Taiwan made
in the wake of the global expansion of capitalism. Thus the lessons learned
could be a lens through which to view other Asian countries' environmental
problems, which have increased dramatically since the 1970s. The ecological
degradation in most of these countries has paralleled economic and industrial
growth. In Taiwan people began to be aware of the problems of pollution and
general exploitation of natural resources only during the 1980s. But it was not
until the 1990s, when the ecological issues attracted global attention, that the
demand for adherence to the international regulations on environmental protection was taken seriously. While Taiwan was hailed as a "great economic success," many from the ecological movement condemned specific traditional social customs, such as consumption of products from wild animals that
undermined wildlife conservation efforts and was hence detrimental to the environment.[1] One observer described the situation as follows:

> Taiwan, one of the "tiger" economies, now has the dubious reputation of being
> the dirtiest place in Asia. The lower reaches of the island's rivers are nearing bi-
> ological death—the result of unregulated dumping of industrial and human
> waste. A third of its rice crop is contaminated with heavy metals and even by Tai-
> wan's lenient standards, air quality is officially harmful for seventeen percent of
> the year.[2]

Some pioneer activists and scholars tried to bring public attention to the
growing problems; however, society at large did not begin to realize the im-
mensity of environmental damage caused by the processes of economic development until the late 1980s. This coincided with the lifting of martial law
in 1987, which was followed by rapid political liberalization and greater social
awareness. During this period social movements cropped up on all fronts involving labor, students, farmers, feminists, and environmentalists.

In the 1990s Taiwan's environmental movement became noticeably
stronger in both the social and political arenas. Nonpolitical as well as politi-

cal forms of environmental activism became widespread, undertaking the tasks of protection, conservation, and preservation of the ecosystem. Furthermore, several Buddhist and Christian institutions became involved in generating environmental consciousness through spiritual and religious perspectives combining moral support and social praxis.

The two organizations studied, both having strong Buddhist ties, have large numbers of women activists and are led by women. In Tzu-Chi women are a majority while in LCA the ratio between men and women is roughly equal. This shows a great concern among women for the preservation of the ecosystem. One recent study on feminist political ecology argues that:

> There are *real*, not imagined, gender differences in experiences of, responsibilities for and interests in "nature" and environments, but these differences are not rooted in biology per se. Rather they derive from the social interpretation of biology and social constructs of gender, which vary by culture, class, race and place and are subject to individual and social change.[3]

In Taiwan women have become more involved in *huan-bao* because of their concern for human-nature relationships. These women feel a special responsibility for protecting the environment and respecting nature. Based on the study of these two groups it is clear that in recent years women have shown a much greater concern and contributed more to environmental activism than ever before. Given these considerations, the first question that should be asked is whether these organizations can be termed "gendered organizations" reflecting a "gendered perspective"? While it is true that a good number of members and volunteers in Tzu-Chi and LCA are women, they cannot be viewed as purely women's organizations, since men also participate in their activities. Also it should be noted that these women are motivated primarily by religious and social reasons, irrespective of, and perhaps largely unaware of, the politics of gender responsibilities toward nature. In the course of this study such issues are analyzed in light of these women's own self perceptions, as reflected in their writings, others' works about them, as well as interviews with them.

Although feminist theory may provide a valuable framework for discussing and comparing the work of these two organizations, the women involved do not draw the main inspiration for their activism from any theoretical base. Rather, their experience determines the theoretical framework in which they (are seen to) operate. In other words, their work is experiential first and theoretical second. Moreover the gap between theory and practice is not as wide as might be expected. Therefore the critical emphasis here is on these women's participation in the "earth-healing praxis." In Buddhist terms this praxis is known as "acting with compassion."[4] As ecofeminist Sally McFague observes, today the agenda of theology and religion has widened to include the context of our habitat—this earth—reflecting a paradigm shift, placing greater focus on praxis.[5]

My focus in this essay is on "feminist action research,"[6] which seeks to enhance the actions of women that directly or indirectly contribute to social and individual change, especially in the lives of women and society in general. Since the data I have collected are not generally known, it is helpful to analyze them here in light of recent developments in Taiwan. Besides these organizations' own Chinese language publications, my sources include interviews, e-mail, and other forms of personal communication with many prominent members from these two groups, including some in leadership positions. In order to understand their position in an increasingly globalized socioreligious activism, these data need to be looked at critically, especially from a theoretically well-informed feminist perspective. However, here I limit myself to a brief historical analysis of these two women-led, religiously motivated, environmental protection groups.

In many ways these women's work breaks down social constructs of gender in traditional Taiwanese society. Women's roles are being redefined each day as the work of these two organizations finds greater appreciation locally as well as internationally. Although I stress maternal imagery in these case studies, I do not intend to create a gender conflict, which could lead to reverse discrimination. For example, some people have asked me, "why only talk about women in discussing environmental protection in Taiwan?" There are some scholars who oppose such an emphasis. Feminist scholar Nancy Frankenberry is one example. In her essay, "The Earth Is Not Our Mother,"[7] she argues that feminists should not perpetuate gendered metaphors in connection with nature, and that ecological issues have to deal with conceptual, political, moral, and practical problems. In principle this is true; however, we cannot ignore the fact that throughout the world women are much more involved in ecological protection activism than men. Frankenberry herself notes that "statistics reveal that women constitute approximately sixty to eighty percent of the membership of most environmental organizations."[8] From the forty-plus interviews I conducted with religious women involved in environmental protection work and related animal rights issues in Taiwan, my general impression is that more women are willing to be volunteers than men. Although I focus on women's contributions, I do not wish to imply that only women can cure the sickness of this earth. The problem is vast and will surely require a great deal of reflection and active cooperation from both sexes, perhaps following the partnership model suggested by Eisler.[9]

In a new social movement such as environmental protection, religious women and their activities are worth documenting. The changes these women have made in society and individuals are significant now and will continue to be so in the future, regardless of where they take place. Therefore I hope my case studies will result in a reappraisal of the role of women in Taiwanese society and generate further interest in their work, allowing women and men alike to draw inspiration from them.

The Tzu-Chi Foundation is a Buddhist organization belonging to one of the major denominations in Taiwanese Buddhism. It plays a major role in Taiwanese religion and grassroots activism within the larger ecological movement. The leader of this foundation, Dharma Master Cheng-Yen, is a Buddhist nun. Of the foundation's over four million members, 80 percent are women, mostly housewives.[10] The majority of the leaders also are women. Among the membership women play a leading role in environmental activities, such as recycling programs, as they realize the importance of self-discipline and thrift. The foundation was selected as the number one social movement in Taiwan in 1991 by *The Global Views Monthly Magazine*, and later awarded the Peace Wind Award (*He Feng*) for its environmental activism and influence in society.[11] Today Tzu-Chi is "the largest civic organization" in Taiwan. In 1994 its worldwide membership of four million included nearly 20 percent of Taiwan's population.[12]

In Taiwan, the story of the founder and the organization's humble beginnings are well known. Master Cheng-Yen started the Tzu-Chi Foundation due to the following two events. The first took place on a hospital floor where she saw a pool of blood on the floor from an aboriginal woman's miscarriage. The hospital had refused to admit her without a deposit, which the aboriginal woman did not have. This shocked and distressed Master Cheng-Yen. The second event was her encounter with three Catholic nuns, who tried to convert her to Christianity. Master Cheng-Yen had a long discussion with them. During their meeting the nuns commented that although Buddhists have the most compassionate doctrine, compared to Catholics they devote little attention to the needs of the poor and the underprivileged in Taiwan. Master Cheng-Yen was deeply affected by this criticism and decided to do something about it.[13]

Master Cheng-Yen founded the Tzu-Chi Foundation in 1966 with thirty followers. They were mostly housewives who contributed as little as two cents a day from their grocery money.[14] Thirty years later, she not only has one of the finest hospitals in the nation, Tzu-Chi General Hospital, but also a training school for nurses, Tzu-Chi Nursing College. Both are dedicated to serving the poor. The foundation has also established the Buddhist Tzu-Chi Free Clinic in the United States, the Tzu-Chi College of Medicine, and a nursing home for the elderly.[15]

In 1985 Tzu-Chi became an international organization when its volunteer work was extended to other countries. In the early 1990s Master Cheng-Yen's followers were spread over five continents and at least nineteen countries. In 1999 that number increased to about twenty-eight countries.[16] According to records published in 1993, the foundation has given away more than $20 million to charity every year.[17] The Tzu-Chi Foundation's work currently includes four major areas of operation: charities, cultural development and publications, education, and healthcare. In addition, the foundation is involved in international relief work, a bone marrow donor program, community services, and a recycling project. These programs are a visible expression of their motto: "one step, eight footprints."[18]

While preaching, the Master uses simple stories rather than profound theories to explain Buddhist doctrine and her ideas of a Pure Land in the human world. She also encourages members to exchange their own stories. Not only are the contents of the stories important, but they also move peoples' hearts. The stories are known to affect their lifestyles, habits, and actions and to invite people to participate in more productive and meaningful activities to continue the Tzu-Chi spirit.[19] Adopting a maternal image, the Master uses a very soft and patient way to change society gradually, to realize the Pure Land step by step. Thus many members contend that when they come to Tzu-Chi, they feel as if they are returning to the comfort of home and are in the arms of a loving mother.[20] These images of the Master and the way she guides people are very different from some Taiwanese Buddhist male leaders, who like to demonstrate their profound intellectual abilities but engage in little social action.

According to Professor Chang Wei-An, it is important to look at the ideals of "this world" Buddhism to understand the social actions of Tzu-Chi members. Master Cheng-Yen often speaks in spiritual language to people suggesting that there is more to Buddhist teachings than what the texts tell us. For example, the Master says, "Do good work in your daily life first, do not go to worship gods in the temple." This implies that the Buddha is alive in this world in one's daily life and is not far away in the heavens or in the temple. The Master also never requires people to give up their normal duties, including their pursuit of their livelihood, to be moral. She rather encourages them to give their best to whatever work they are involved in.[21] Master Cheng-Yen urges her followers to "perfect their regular work" before becoming members of Tzu-Chi. Once someone asked her how a housekeeper could become a member of her organization. Her answer was that the first step was to become a good housekeeper and a kind mother.[22]

Master Cheng-Yen regards environmental protection as connected to the mind, health, and the earth. "Environmental protection must start from the mind, . . . if everybody can get rid of greed, anger, delusion and pride, then all people can help each other and work together to open up a piece of clean land."[23] Further, she says, "Tzu-Chi's beauty lies in the conviction that in the world there is no one I cannot love, no one I cannot forgive and no one I cannot trust. When all minds have been purified, then all hearts can be linked together."[24] Once she quoted from the lines of a song, "The Gardener of the Great Earth," which describes the sadness of mountains bringing tears to one's eyes:

> The sky is broken
> Who would not be saddened?
> The green mountains are shedding tears.
> Who would not be sorrowful?[25]

Here she stresses that "we have no claim to our bodies, but only the right to use them." To use them properly requires wisdom. "If we live with the concepts of environmental protection for mind, health and the great earth, our society will be rich and we will always be lucky and happy."[26]

Master Cheng-Yen urges all people to pursue a simple life or to change their daily lifestyle. "After we have gotten rid of greed, anger, delusion, arrogance and suspicion, then we must purify our bodies. In other words, we must do environmental protection for health."[27] She also asks them to cherish the things they have (*fu*, which literally means "good fortune").

> Environmental protection for health is very important. Life is so short that we must look after our health well. Otherwise, instead of making good use of our life, we have to spend a lot of time seeing doctors and taking medicine. That would be a great waste. We don't need to be too fussy about the quality of our clothing, food and housing. The most important thing is whether they are clean and sanitary.[28]

Master Cheng-Yen emphasizes action to change the situation, making amendments in one's daily practice, and working with the entire community. For example, resource recycling requires cooperation from different communities. These communities may not be located in the same city or village. Through various activities that enrich their lives they try to build community consciousness. This kind of community is not bound by geography, but rather by a set of principles and certain praxis that can be called an "invisible community."[29] Thus Tzu-Chi activism is not confined to individuals; their activism is also carried out at the organizational level. It is both a social and a religious activity. Tzu-Chi offers a spiritual location within a social geography for people to purify their hearts and enter into another sacred world.

Master Cheng-Yen lives in the "Abode of Still Thoughts," the spiritual home of Tzu-Chi members,[30] and her spirit of environmental protection is carried out constantly. She often says that the earth's resources are right under our feet. The Pure Land is not something far away but right here around us. Therefore, "the resident nuns do not use pesticides on the vegetable garden at the Abode.... [They] collected wood blocks from the mountains, wood chips from wood shops and peanut shells from cooking oil shops as fuel for cooking."[31] They urge people to use every part, recycle everything, and waste nothing.

> The nuns use natural detergents like soybean powder to wash pots and dishes, so there is no need to worry about harmful residue from chemical detergents. In addition, after the dishes are washed, the powder sediment can be used as fertilizer and the water can be used to water the crops. Every part is used and nothing is wasted.[32]

The September 1999 earthquake (7.3 on the Richter scale) left more than a thousand people dead and many more homeless. Among other things, approximately eight hundred school buildings were destroyed. As one interviewee recalled, Master Cheng-Yen refused to call the earthquake "nature's revolt," as some did. Instead she explained that "Earth is our mother and, while all this time she tolerated our mistreatment of her resources, there is always an end to her endurance and this time she gave in. Now we must intuitively comprehend mother's heart and realize our responsibilities towards her."[33] Thus Master Cheng-Yen encouraged her followers to direct all their compassion toward those affected by the disaster. As a result of this call from the Master, Tzu-Chi received donations of $168,169,714. Roughly half of that money was received within ten days of the disaster. This was viewed as a miracle and also attested to the level of trust that Tzu-Chi has established among its followers and beyond.[34]

Master Cheng-Yen hopes that all members of Tzu-Chi will begin from their own homes and extend their love to include the "great earth," the dwelling place of all. Therefore, beginning in 1990, she asked that everyone recycle and at the same time generate less garbage. "We produce garbage therefore cleaning the great earth is also our responsibility."[35] The members immediately organized environmental protection volunteer teams in different cities and villages in Taiwan to start the process. Other people were influenced by them and joined in the work, increasing the number of people involved in recycling dramatically.[36]

By undertaking these practical activities, the members are able to relate to as well as live the "organic connection"[37] between Buddhist ideology and environmental protection. Praxis in the form of recycling also enriches the conceptual basis of the theory. For example, previously those responsible for collecting resources were only able to speak of the praxis in Buddhism, but they were not able to demonstrate it. Now, after many years of experience, they are able to speak systematically about environmental protection within the framework of Buddhism.[38]

In 1994 the money generated from recycling was $1,678,570.[39] In 1995 recycled paper saved 366,752 trees, which is equal to a small forest.[40] In 1999 the total amount of paper recycled was 52,011,931 kg. If 50 kg of paper saves one twenty-year-old tree, then in 1999 alone, Tzu-Chi's recycling efforts saved an equivalent of 3,450,079 trees.[41] Most of this money is then used for charity works, such as funding hospital and school buildings and helping victims of natural disasters. Thus Tzu-Chi members believe they are living their slogan: "Turn garbage into gold; turn gold into a loving heart."

Taiwanese women tend to stress the nurturing aspect of nature. The focus on nurturing allows them to see the earth both as a mother and a sick child. In traditional Chinese society, women were required to have children, for they were seen as responsible for the next generation. In contemporary Taiwanese society, women are not limited by the tradition of having and nurturing children. In the 1990s greater numbers of women participated in the Home Makers' Union and Kaohsiung Huan-Bao Mother Service Team. Women's groups tend to emphasize household over lineage, raising broad themes of nurturance for children and for the earth.[42] The preservation and improvement of the environment, in the broadest sense of the term, is not seen as outside the scope of maternal duties.

Tzu-Chi also quite typically downplays philosophical subtlety and does not urge people to take monastic vows. It also differs from earlier movements by focusing clearly on the idea of charity as a way of improving *karma*.[43] Many middle-class women extend their family values and roles to the wider society by becoming community leaders. One interviewee related a story of her transformation:

> I am only a high school graduate and have always been active in environmental work, but once I was asked to speak in front of an audience that included principals from various schools and educators, government officers and prominent journalists. Since this was my first time speaking in public, I felt very nervous and did not know how to address them, so I wrote a two-page speech. Somehow I managed to read the first page, but due to nervousness I could not turn the page over to read the second page. After a brief pause, one female journalist realized my difficulty and came to my rescue, turning the page for me. This was my first experience. Now I comfortably speak in front of large audiences of all kinds, including military gatherings.[44]

In Tzu-Chi there is another important relationship being highlighted from a more cultural perspective, the emphasis on family values. Significant is the fact that many of the *huan-bao* activities conducted by Tzu-Chi are family-oriented. In some ways if one member of a household is connected, she or he is encouraged to get other members enthusiastic about environmental work as well. One interviewee spoke of how the family of another volunteer was saved from depression by getting involved in the relief work after the September 1999 earthquake. They found that their own little worries in life were nothing compared to the real problems of those affected by the disaster. Even after the relief work was over, the entire family continued their involvement in the recycling project.[45]

Professor Yong Hui-Nan, a famous scholar in Taiwan, criticizes Master Cheng-Yen's environmental movement as focusing mainly on recycling and planting trees in local contexts only. The Master also emphasizes that environmental protection must focus on getting rid of spiritual garbage (e.g., greed, anger, delusion, and pride) from within. These things, in the eyes of Professor Yong, are attempts to bring about internal (moral and spiritual) change rather than external change, in other words structural change in society and the polity. Therefore Master Cheng-Yen's concept of environmental protection is insufficient, because it is limited to a few problems from the point of view of the ecological crisis and does not address issues such as industrial and nuclear waste.[46] In addition, Professor Yong also remarks that members of the Tzu-Chi Foundation are not involved in any political activism against government policies, which are responsible for creating many environmental hazards.[47] Therefore Professor Yong suggests "a type of Buddhist ecology which would emphasize 'inside spiritual mind' and 'outside objective environment.'"[48] In other words, he would like to see attention paid to both the internal and external aspects of the environmental movement.

Professor Yong's critique is quite valid, although due to his being male some tend to think that he reflects a male bias against the feminine values inculcated by Tzu-Chi. However this is not the case, as he also criticized a man-led, Buddhist organization, Dharma Drum Mountain, for its nonpolitical stance. Rather I feel that Yong's approach is based on liberal and radical standards, in terms of which he judges Tzu-Chi to be a conservative organization. Therefore some radical Buddhists strongly urge for a change in public policy to avoid suffering at the individual and also at the communal level. I think that the above perspective is acceptable because Buddhist *metta*, loving-kindness, implies not only compassion but also action, which is required for all communities in order to change an inadequate public system and thus improve people's mutual *karma*.

It may be unfair to expect that Tzu-Chi could fulfill every function for protecting the environment. Since most of the members are women, they are already busy taking care of their families as well as their careers. Thus it may be difficult for Tzu-Chi members to deal with external change as well as internal transformation. The criticism that they are not dealing with both these aspects of the environmental movement is perhaps out of place, for they are doing as much as is humanly possible, while still growing at a remarkable pace.

Master Cheng-Yen and members of Tzu-Chi are engaged in the meaningful work of transforming religious visions and women's roles in Taiwanese society. For thousands of years Buddhist ethics have traditionally emphasized behavioral guidelines and liberation for the individual from within rather than structural changes from without. This can make it very difficult for Buddhist religious or social leaders to advocate social change. But the Tzu-Chi Foundation offers and establishes a social network that can be mobilized, that is mass-based. They emphasize that religion can be instrumental in bringing about social change. It is noteworthy that women instead of men now play leading roles in different religious groups and organizations. These women provide the main channels through which leaders can normally mobilize local people. It is especially unusual to have an organization of this scale in Chinese culture. But as Master Cheng-Yen says, "every one of us can be a gardener of the great earth. I hope all of us will take action to save the mountains, recycle resources and clean the earth."[49] Taiwan is considered one of the four "Asian Dragons" based on its fast economic rise. Even though this so-called economic miracle brought some material comfort, it can never compensate for people's loss of clean air and water.

Upon examining the Taiwanese Buddhist thinking of the last fifty years, we find that the awareness of Buddhism is changing and developing constantly. In recent years, ecological consciousness has increased among many believers. According to Professor Jiang Tsan-Teng, Taiwanese Buddhism has two main streams, one conservative and the other revolutionary. The conservatives put more emphasis on *nirvana* and focus on the worship of Amitabha's Pure Land as part of their central religious beliefs. The corresponding performance of religious ethics includes *Hu Sheng* (protection of life) and *Fu* (cherishing *fu*, which literally means "good fortune").[50] The so-called revolutionary group, on the other hand, has been instrumental in generating interest in environmental work. From this very group came several important environmental activists and leaders, for example, Chuan-Dao, Chao-Fei, Yong Hui-Nan, and others.[51] They are all very active in the environmental protection movement, although they have different emphases. Among these three leaders, Chao-Fei is the only woman.

Sakya Chao-Fei, a nun, founded the organization called Life Conservationist Association (LCA) in 1993. Since its inception, LCA has contributed to the cause of environmental protection in much more novel ways than other such movements. LCA members are trying to push Taiwanese society to change its trends and ways of life by demanding systemic change through government regulations. They also have been involved in raising awareness about practices that have implications for the environmental and conservation issues in Taiwan. For example, Chao-Fei led a large protest against the exploitive fishing practices and the abuse of horses in horse racing. This generated quite a discussion, because this issue concerned "the state of non-human beings," which had not been a priority issue in Taiwanese society previously.[52]

For Sakya Chao-Fei, the starting point for environmental protection was her protest against "fishing without bait" (*fan cuo yu*), which started in 1992. At that time there were many places in Taiwan which organized *cuo yu*, a fishing game that involved the use of a special curved hook to catch fish in indoor fishing ponds. Sakya Chao-Fei felt that this game was very disrespectful of life and amounted to mistreatment of fish. The game was so popular that it was even encouraged by some parents as a means of entertainment for children without considering the inherent abuse of life and resources.

Sakya Chao-Fei found support among some Buddhist masters and other influential individuals, who came together to protest against this cruel fishing game. She organized press conferences, which highlighted the plight of fish, and registered protests against this practice. From then on she became more involved in improving the situation of animals in Taiwan. She felt that large number of animals were tortured by humans, and that humans should change the way they treated animals. Thus Chao-Fei started the first social pressure group that advocated animal rights in Taiwan. She wrote popular articles concerning the human-animal relationship, redefining the way humans have traditionally treated animals. Sakya Chao-Fei's actions in support of animal rights represented "the first public expression of concern for 'non-human beings' in contemporary Taiwan, ... initiated from a religious perspective."[53]

In the process of conducting the anti-*cuo yu* movement, she found that without the law backing the social movement, its influence would not last. For example, those who run these *cuo yu* businesses were engrossed in their profit-making, which sometimes exceeded several million dollars per month. If someone asked them to follow the moral path of the heart, they were unlikely to agree because of the large profits involved. When Sakya Chao-Fei confronted the legal issue and policy makers implemented a minimal penalty on those businesses for abusing animals, the psychological effect worked to eliminate *cuo yu*. From this Chao-Fei gained experience in utilizing the legal process to bring about the successful implementation of her principles of environmental protection.

In 1994 the government invited Chao-Fei to join a committee, which eventually helped ensure the passage of a bill to protect animals. At that time, a business group that sought to develop a horse-racing track had been given permission by the government of Taipei, the capital city. Sakya Chao-Fei was strongly against horse racing, because she believed it involved cruel treatment of horses during their training. An average horse had 70 to 80 percent of its bones broken. In addition, the medicines used in training created internal bleeding. If the horse did not perform up to the requirements and failed to attract betting, it was immediately killed. In order to maintain the dignity of the horses, she fought to keep racing from becoming legal. Thus the entire horse-racing lobby was defeated by a group of "unprofessional" nuns and monks who inspired people's consciences.[54]

Chao-Fei has a number of publications on the theory and practice of Buddhism. She also regularly contributes articles to magazines and newspapers promoting environmental protection. Her long-running columns appear in two newspapers in northern Taiwan, *Min Jong Daily News* and *Taiwan Times*.[55] Besides writing for magazines and newspapers, she has published many books, some of which are related to environmental issues.

According to Sakya Chao-Fei, there are two extremely important doctrines in Buddhism. One is the law of *yuan qi* (Pali, *paticca samuppada*) or dependent co-arising. Buddhists believe that all beings are dependent on each other and nothing exists by itself. This is the notion of "interbeing," in which all things exist in relation to each other as part of an interconnected wholeness.[56] Hence all beings are equal. The second concept, *fu sheng* or "protecting life," is the ultimate spirit of Buddhist ethics. Since all beings are interconnected, if one being is hurt all beings are affected directly or indirectly. Therefore love and kindness for others should arise in one's heart naturally.[57]

Sakya Chao-Fei's *Buddhist Ethics* has popularized the phrase "all beings are equal" (*zhong sheng ping deng*), so much so that it has become an important slogan for many animal rights activists. Chao-Fei asserts that every life is of equal value. There is no basis for claiming that an inferior life should serve a superior life. This hierarchical consciousness cannot serve as a moral ground but only exists to justify the strong destroying the weak in the ecological scheme. Therefore Chao-Fei has devised three ways to save animals. First, try to move peoples' hearts by emotional appeal in favor of animals. Second, educate people by inculcating rational thinking. Third, demand legislation to protect animals. In other words, people need to have various channels of motivation in order to develop compassion.

This is the spirit of sharing the same crises and the same benefits. This approach is much better than developing personal virtue and is consistent with the doctrine of *wu wo*, which implies "no-self."[58] Chao-Fei also argues that if a society is cruel toward its animals, it will tend to practice cruelty toward humans as well.[59]

According to Sakya Chao-Fei, one should not expect that everyone's choices will be one hundred percent perfect, but when people reflect on choosing a middle way they will ultimately find it. In other words, in different situations one must opt for the best choice to solve the problem. Although it may not be absolutely right, it may be the best available in a given situation. To make such decisions requires wisdom. If people are not selfishly driven, Chao-Fei argues, then they are more likely to keep the wisdom of the middle way. For example, should you save the cow or the wandering dog if both are bleeding on the road? Should you raise a tiger, which is an endangered species, if by raising it you endanger the lives of many cows? These are difficult questions, but we cannot refuse to act because we do not have absolute answers. We have to choose our path in the current situation to find a way that is better than the others, so long as that does not make the situation worse.[60]

If all beings are equal then we face the question: Why do humans kill animals to eat them? Sakya Chao-Fei knows not everyone can be a vegetarian. Sometimes animals are needed to satisfy our appetite or for economic benefit. The situation or need may be unavoidable, but one still should not take it for granted and console oneself by thinking that this is the fate of that animal. "We should feel shameful and appreciative at the same time while eating animals or making money from them, because any animal, regardless of its species, would love to be alive. If we cannot make them live anymore, at least we do not need to let them die so dramatically or so frivolously."[61] If people ignore the suffering of animals, they can also easily ignore the suffering of domestic servants, women, and aboriginal people because they all have less power.

Although LCA is not a Buddhist organization as such, most of its members are Buddhists. The remainder are Christians and experts in environmental issues with no connection to any religion. They all have a common concern, animal protection. In January 1993, Sakya Chao-Fei, Wu-Hong (a Buddhist monk), Reverend Lu Jun-Yi (a Protestant minister), and Father Wang Jing-Hong (a Catholic priest) formally launched the LCA organization by joining forces.[62] The organization has three major aims: (1) to educate the public on how to love and protect animals; (2) to pursue legal channels to bring about change through government; and (3) to prevent activism from causing a social crisis.

One method for achieving this is through publishing books, magazines, and other teaching materials to educate the Taiwanese. Chao-Fei, the founder, has written many books and articles. Since its inception, LCA has also published an official quarterly newsletter, *Animals' Voice of Taiwan*. With a circulation of twenty thousand, it is on its twenty-third issue already. The newsletter is widely distributed through major bookstores in various cities in Taiwan, as well as through the mail.[63] It is given out free of charge in order to create greater awareness of the activities of the animal rights movements in Taiwan and the rest of the world. They have also produced four educational videos about animals used in the economy, about stray dogs, and a documentary on Taiwan's animals. Other publications include a dozen or so books and manuals on ways to protect animals, including how to take care of or sterilize stray dogs.[64]

LCA has also organized exhibits and lectures in many cities and towns on subjects such as "Reject Eating, Buying and Raising Wild Animals." They also sponsored a family fair demonstrating ways of caring for and protecting animals in the Taipei Zoo.[65] Volunteers are recruited to run, as well as participate in, summer camps for children where these issues are discussed.

By using organizational power to form a pressure group to promote animals' legal rights and also by organizing and negotiating, the abuse of animals can be prevented. This emphasis on making laws to fight for animal rights started in 1993.

From the preceding it is not difficult to ascertain that LCA went to great lengths to attach social and political substance to the moral issue of animal protection. During my interview with Sakya Chao-Fei, I asked her what she considered her greatest achievements as an activist. She mentioned the following two activities. In September and October 1997, LCA campaigned against a visiting "Great European Circus," because they believe it is unethical to use animals for entertainment purposes. These animals often suffer at the hands of their trainers and owners. Although major corporations spent a great deal of money to promote and prepare for the circus, in the end very few people attended. LCA received overwhelming support from other social organizations both in Taiwan and overseas.[66] Sakya Chao-Fei's second major achievement was also in 1997, when the Taiwanese legislature passed laws banning horse racing. In fact, the law banned misuse of all kinds of animals, including horses and dogs, for the purpose of gambling. Thus Taiwan became the first country in the region to have such laws in support of the animal rights movement.[67] Chao-Fei credits her co-workers, activists from other social movements, and academics, as well as the media, for these important milestones in LCA's history.[68] Today LCA continues to be a watchdog organization monitoring the implementation of laws for animal protection.

In Taipei, at the Dr. Sun Yat-Sen Memorial Hall, there is a beautiful pond called *cui hu*, which means the "bluish-green lake." Due to some repairs to the pond, the fish living in the pond had to be removed. The authorities decided to hold an open "fun contest for parents and kids" by allowing them to get in the pond and catch the fish by hand. This supposedly "fun contest" was to be held on June 18, 2000, and three hundred participants were to be allowed inside the pond. Two days before the event, LCA made a public announcement through the media asking people to refrain from this "fun" activity for the sake of the fish. They argued that it amounted to cruel treatment of these living creatures and was a bad example for the younger generation.

According to Sakya Chuan-Fa, a Buddhist nun and the current secretary general of LCA, the lack of professional training in handling the fish, as well as reference to the activity as a "contest," reflected insensitivity toward these aquatic creatures. LCA requested that the Memorial Hall not go through with this program. In response, the Memorial Hall simply changed the title of the activity, calling it, "Loving *Cui Hu*: Adopting and Caring for Aquatic Animals." According to LCA this was simply a change in form but not in content, so they decided to protest in front of the Memorial Hall on the day of the activity.[69] Eventually, after much discussion, the Memorial Hall agreed to drop the idea of turning the task into a public activity. A few professionals removed the fish, then gave them to families who wished to care for them at home. This incident gave LCA and its animal protection activities great exposure. It also made people more sensitive to other living beings, while contributing greatly to building LCA's reputation in animal rights issues.[70]

Sakya Chao-Fei believes that a social movement is an instrument of change. She asks that we correct the wrong concept first, and then go further to change the behavior. This kind of movement conflicts with those people who do not want any change. Therefore the important point for the movement is to learn how to minimize potential opposition. LCA's motivation is not to convert throngs of people to Buddhism, rather it seeks to save animals. "If there is any way to accomplish this, then we will use that way and we will not use overly religious ways to pursue it," states Chao-Fei.[71] For this reason it may be noted that people of different religions are involved in LCA, because they have the same goal, to promote social concern and to work for social change.

LCA represents, among other things, a front for the humane treatment of animals in Taiwan. They are steadfast in their mission, even though they are often questioned about the wisdom of being attentive to the question of animal rights when an even a greater problem exists with human rights, which perhaps should be given priority. But for Sakya Chao-Fei human rights and animal rights are not in conflict and are not mutually exclusive. They can be worked on simultaneously; first, because human civilization can be raised from the level of "survival of the fittest" to compassion for humanity's sake, where practicing human rights is moral common sense. The weaker have the right to survive, and they should not be deprived of their dignity. If this is so, animals' rights should have the same emphasis as human rights, because animals cannot speak for themselves. They are the weakest among the weak.

Second, an emotional fluctuation exists in peoples' behavior, and the gap between love and hate is very wide. This type of thinking is easily turned against itself. It is far removed from human and animal relationships, which are simple and pure. Therefore, to cultivate humans' sympathy for "loving the weak," making animal rights an issue will be helpful because this is the extended heart of *Ren* (humaneness), and it indirectly improves human rights. In other words, Sakya Chao-Fei argues that speaking for the weakest of the weak, that is, animals, does not imply putting emphasis on animal rights over and against human rights. Rather it challenges an overemphasized anthropocentrism.[72] However, it is important to remember that Chao-Fei's emphasis on animal protection is not the same as the Western understanding of animal rights, which according to her is approached from the perspective of metaphysics and ethical obligation. In contrast, Chao-Fei draws her inspiration from the Buddhist notion of "dependent co-arising," which implies that all beings are equal.[73]

In traditional Chinese culture certain days of each month were designated for not killing any animals. These were known as "vegetable eating days" (*chi su*). Similarly, during certain festivals people would set animals free (*fang sheng*). These two customs are still practiced today. But Sakya Chao-Fei thinks that in modern times Buddhism can do more by being a key element in influencing public policy in favor of animal protection. Chao-Fei states, "we should not easily give up this right and obligation; we should not consider all protests and public action as something terrible." She commented that most successful social activities have been carried out with the help of non-Buddhists—Christians and others—who are an integral part of the social movement, while some other Buddhist groups for the most part have kept their distance.[74] In some ways Sakya Chao-Fei and LCA are trying to compensate for the inaction of some Buddhists in Taiwan. She has already had a tremendous impact on public policy, which is more than can be said for those who adhere to traditional modes of Buddhist involvement in social life.

So far LCA has helped bring about three major policy changes regarding animal protection in Taiwan: laws protecting wild animals, laws protecting domestic animals, and the law against horse racing. Without LCA's efforts animals in Taiwan today would suffer much more than they do. There would also be a greater number of social problems. Finally, linking their action-oriented life to theory, Sakya Chao-Fei observes that the "collective *karma*" (*gong ye*) can be transferred into a good direction through gathering "collective wishing" (*gong yun*).

Once feminist theologian Katie Cannon asked me, "do you believe that religion can really change society?" At that time I was not so sure, but now I can say that from the perspective of the cases studied here, the answer is absolutely yes. Religion *can* bring about meaningful change in society, even though it may be manifested in many different ways.

The founders and other leaders of both of these groups are nuns who have earned great respect from the wider Taiwanese society for their social work. They have changed the image of both the female's role and religion's role in Chinese society. Even today in some countries Buddhist nuns still cannot be ordained, such as in Thailand, Myanmar, Sri Lanka, and Tibet.[75] But the Buddhist nuns from Tzu-Chi and LCA have helped elevate the role of women in general and religious women in particular.[76]

Through this environmental action these women are not only engaged in social activism and spiritual enlightenment, they are also boldly redefining their identities; they are contesting the meanings of gender in Taiwanese society. This gender transformation has begun to redefine environmental activism itself, which now includes women's, especially religious-minded women's, knowledge, their experiences, and their particular modes of operation.

Here the question of self-designation is also very important: do these women call themselves ecofeminists? From the interviews that I conducted, I found many of them are hesitant to be associated with feminists. Some admitted that they are not familiar with the term "*ecofeminism*." So in which category do they belong? Most of them would certainly be called "environmental activists," even though these groups are involved in much more than just environmental issues, especially LCA. But can they be categorized as ecofeminists? This depends perhaps upon how the term is defined. Such a categorization would not be an imposition if we define ecofeminism as that which encourages women to change their lifestyles. If ecofeminism is truly related to the practical in addition to the political, as argued by some,[77] then many of these women would qualify as ecofeminists. Nevertheless, these women are not working for environmental protection as feminists. Thus we can say that the work of these organizations is neither purely political, nor purely feminist.

Of the two groups Tzu-Chi has a greater concern for making the connection between spirituality and environmental protection. Thus environmental protection is seen as a means to an end, which is spiritual merit. In contrast, LCA's main goal is to rethink the relationship between humans and other living beings and to encourage political authorities to make policy changes that will effect social change. Their social movement is an end in itself.

Tzu-Chi is a silent social movement, as it places a greater emphasis on individuals to change their lifestyle and seek enlightenment by removing the spiritual garbage from their hearts. LCA, on the other hand, is an aggressive and politically active movement. It cooperates and connects with other similar groups working for social change. Tzu-Chi focuses on "cherishing *fu*" and "cherishing *wu*." *Fu* implies the preservation of spiritual and material resources, including things like love and grace, while *wu* implies the actual material things that people possess. LCA focuses on protecting animals, asserting that every being is equal.

Even though Tzu-Chi tries to bring about social change from below, at the individual level, its members have a strong collective identity as a group. Basically they are against protests and demonstrations on the street. However, their organization's mobilizing techniques are well developed, and as a result they have over twenty thousand volunteers who engage in various activities linked to environmental protection in their respective communities, such as recycling and planting trees. They see themselves as working independently, but spiritually and institutionally they are affiliated with the collective Tzu-Chi organization. Among the main connections between these communities are their contributions to Tzu-Chi via local recycling units. They feel that when they are engaged in *huan-bao*, it is primarily for ethical and moral reasons. The recycling project generates money, which is then sent to Tzu-Chi's headquarters in Hua Lian to be used for other social projects.

LCA seeks to bring about structural changes by applying pressure on political leadership, which will then make fundamental policy changes through laws affecting society. Their method is somewhat confrontational and includes protesting against those governmental policies that are not environmentally sound. They try to inculcate awareness of the value of life of all beings. They effectively utilize the media to spread their message among the masses to invoke social responsibility (as opposed to moral and spiritual concerns as in Tzu-Chi). Compared to Tzu-Chi, LCA is a much smaller group of approximately 350 members total, as mentioned in their 1999 Working Report. In the past LCA played an important role as a pressure group leading a social movement. In recent years they have become more involved in cooperating with local governments and helping in the process of creating a legal basis for the application of environmentally beneficial policies.

Tzu-Chi is an organization that worked alone from the start; they have no confusion about their being a religious, Buddhist organization. LCA, on the other hand, has a somewhat open status and does not present itself as a religious or particularly Buddhist organization. LCA often invites other people from different groups, religious and otherwise, and seeks cooperation from a wider social base. Hence one may find many non-Buddhist and even nonreligious people in LCA's list of members. But the founder and leadership of LCA are Buddhist nuns, hence its characterization as a Buddhist group.

Both groups are good at publicizing their efforts and activities through an effective use of print and electronic media, making information accessible to larger audiences. In some ways, LCA is working on all fronts, realizing that no single variable in a given society can cause meaningful change. Thus they attempt to mobilize both formal and informal structures in Taiwanese society by linking the state, organizations, community, and of course the individual. Tzu-Chi focuses on the individual and the family unit, which is seen as the foundational link to all the rest of the variables. Moreover, by linking religious truth to the issues of *huan-bao*, by default Tzu-Chi's struggles become very personal and at the most communal in their primary motivation.

In conclusion it may be said that in Taiwan, women's contribution to the cause of environmental protection is very valuable and noteworthy. It has affected all levels of human activity. It is also clear that as Taiwanese religious women involve themselves in the environmental protection movement, they experience new possibilities for development in terms of spiritual reform, individual lifestyle change, reorganization of human relationships, action-oriented politics, communal solidarity, effective media operations, and greater interreligious understanding. All these are by-products of these women's efforts in one social movement that also greatly contributes to ecofeminist activism.

Taiwanese religious women involved in *huan-bao* are very different from the radical ecofeminists in the West, who define the problem primarily in terms of androcentrism and hierarchical dualism. Taiwanese women insist neither on "delinking masculinity as power" as an instrument of social change, nor on "social redesign on feminist principles."[78] Rather they are more in tune with the ecofeminist Karen Warren's idea of "tranformative feminism,"[79] which emphasizes a holistic treatment of issues. The diversity of these women's experience in Taiwan should also be appreciated, as they present their experimental praxis, lifestyle improvement, storytelling methods, and sociopolitical structural reform in conducting *huan-bao*. However, it is noteworthy that Taiwanese religious women in grassroots movements have a unique aspect that allows them to consistently link themselves with their religio-cultural commitment and communal solidarity. This includes family involvement, as well as interreligious cooperation.

And finally, to sum up these religious women's environmental agenda, I quote Nelle Morton: "The spiritual is experienced profoundly as sisterhood in its loftiest and most universal sense and, we may add redundantly, political action of the most radical sort on behalf of and ultimately including all humanity—women, children and men."[80] And, I may add, animals and plants as well!

Notes

1. Hua-Bi Tseng, "The Shaping and the Significance of Environmental Movement in Taiwan. Part 3: The Transformation of Ecological Thoughts in the Period from 1970s to 1990s" (Report prepared for the National Science Council, 1996), 4.

2. Tricia Caswell, "Australia and Asia—The Environmental Challenge," in *Living with Dragons: Australia Confronts Its Asian Destiny*, ed. Greg Sheridan (St. Leonard's, New South Wales: Allen and Unwin, 1995), 69.

3. Dianne Rocheleau, Barbara Thomas-Slayter, and Esther Wangari, eds., *Feminist Political Ecology: Global Issues and Local Experiences* (London and New York: Routledge, 1996), 3.

4. Stephanie Kaza, "Acting with Compassion: Buddhism, Feminism, and the Environmental Crisis," in *Ecofeminism and the Sacred*, ed. Carol J. Adams (New York: Continuum, 1994), 50ff.

5. Sally McFague, "An Earthly Theological Agenda," in *Ecofeminism and the Sacred*, ed. Carol J. Adams (New York: Continuum, 1994), 88.

6. Shulamit Reinharz, *Feminist Methods in Social Research* (New York: Oxford University Press, 1992), 175–96.

7. Nancy Frankenberry, "The Earth Is Not Our Mother: Ecological Responsibility and Feminist Theory," in *Religious Experience and Ecological Responsibility*, ed. Donald Crosby and Charley Hardwick (New York: Peter Lang, 1996), 23.

8. Frankenberry, "The Earth Is Not Our Mother," 30

9. Riane Eisler, "The Gaia Tradition and the Partnership Future: An Ecofeminist Manifesto," in *Reweaving the World: The Emergence of Ecofeminism*, ed. Irene Diamond and Gloria Feman Orenstein (San Francisco: Sierra Club Books, 1990).

10. Chien-Yu Julia Huang and Robert Weller, "Merit and Mothering: Women and Social Welfare in Taiwanese Buddhism," *The Journal of Asian Studies* 57, 2 (May 1998): 379.

11. Wei-An Chang, "Buddhist Tzu Chi and Recycling Resource," in *Buddhism and Social Concern—Life, Ecology and Environmental Concern*, ed. Chuan-Dao (Taipei: Modern Buddhists Association, 1994), 71.

12. Huang and Weller, "Merit and Mothering," 379.

13. Heng-Ching, *Wonderful Women in the Bodhisattva's Way* (Taipei: Dong Da, 1995), 180.

14. *Ten Thousand Lotus Blossoms of the Heart: Dharma Master Cheng Yen and the Tzu Chi World*, Pamphlet (Tzu-Chi Foundation), 12–13.

15. Barry Corbin, John Trites, and Jim Taylor, "Religious Leaders Making a Difference," in *Global Connection: Geography for the 21st Century* (Toronto: Oxford University Press, 2000), 408.

16. Tzu-Chi Foundation, August 16, 2000, <www.tzuchi.org.tw>.

17. Huang and Weller, "Merit and Mothering," 379.

18. Tzu-Chi Foundation, August 16, 2000, <www.tzuchi.org.tw>.

19. Wei-An Chang, "Silent Social Reform: The Compassion Relief Tzu-Chi Foundation as a Model for Social Change," paper presented at the Annual Meeting of the American Sociological Association, San Francisco, 1998, 17–18.

20. Hwei-Syin Lu, "Gender, Family and Buddhism—Tzu-Chi Foundation as an Example," in *Gender, Spirituality and Taiwanese Religions*, ed. F. Li and R. Ju (Taipei: Academia Sinica, 1997), 108.

21. Chang, "Silent Social Reform," 14.

22. Shyan-Ming Chao, *Three Taiwanese Giants* (Taipei: Kai Jin 1994), 15.

23. Cheng-Yen, "Let's Do It Together—Environmental Protection for Mind, Health and the Great Earth," trans. Norman Yuan, *Tzu-Chi Quarterly* (winter 1996): 8.

24. Cheng-Yen, "Let's Do It Together," 8.

25. Cheng-Yen, "Let's Do It Together," 9.
26. Cheng-Yen, "Let's Do It Together," 7–9.
27. Cheng-Yen, "Let's Do It Together," 7–9.
28. Cheng-Yen, "Let's Do It Together," 7–9.
29. Chang, "Silent Social Reform," 78.
30. Shun-Yen Chan, "Environmental Protection at the Abode of Still Thought," trans. Tracy Tai, *Tzu Chi Quarterly* (summer 1997): 24.
31. Chan, "Environmental Protection at the Abode of Still Thought," 24–25.
32. Chan, "Environmental Protection at the Abode of Still Thought," 25.
33. Wu Jin-Sz, interview by author, June 30, 2000.
34. *Tzu-Chi Dao Lu Newsletter* 341 (April 2000).
35. Shan-Jun Gan, "Hand by Hand Heart by Heart Garbage Became Gold," *Tzu-Chi Yearbook* (1995): 3.
36. Gan, "Hand by Hand Heart by Heart Garbage Became Gold," 2.
37. Chang, "Silent Social Reform," 78.
38. Chang, "Silent Social Reform," 84.
39. Gan, "Hand by Hand Heart by Heart Garbage Became Gold," 3.
40. Gan, "Hand by Hand Heart by Heart Garbage Became Gold," 5.
41. Tzu-Chi Foundation, August 16, 2000, <www.tzuchi.org.tw>.
42. Hwei-Syin Lu, "Women's Self-Growth Groups and Empowerment of the 'Uterine Family' in Taiwan," *Bulletin of the Institute of Ethnology* no. 71 (1991): 34ff.
43. Chang, "Silent Social Reform," 80–81.
44. Yang Fen, interview by author, June 23, 2000.
45. Lin Siu-Hua, interview by author, July 18, 2000.
46. Hui-Nan Yong, "Review of Contemporary Taiwanese Buddhist Huan-Bao Belief: Cases from Prescribing Pure Land on Earth and Spiritual Huan-Bao," *Contemporary World* 104 (1994): 32ff.
47. Yong, "Review of Contemporary Taiwanese Buddhist Huan-Bao Belief," 42.
48. Yong, "Review of Contemporary Taiwanese Buddhist Huan-Bao Belief," 45; Chang, "Silent Social Reform," 8.
49. Cheng-Yen, "Let's Do It Together," 9.
50. Yih-Ren Lin, The Environmental Beliefs and Practices (Ph.D. diss., University College, London, 1999), 193.
51. Tsan-Teng Jiang, *Contemporary Taiwanese Buddhism* (Mandarin) (Taipei: Nan Tian, 1997), 109–10.
52. Lin, The Environmental Beliefs and Practices, 245.
53. Lin, The Environmental Beliefs and Practices, 234.
54. Chao-Fei, *Although Tired Birds Still Fly into the Sky*, trans. Wan-Li Ho (Taipei: Fa Jie, 1996), 172–75.
55. Chao-Fei, *Tired Birds Still Fly*, 186.
56. Bill Devall, "Ecocentric Sangha," in *Dharma Gaia—A Harvest: Essays in Buddhism and Ecology*, ed. A. H. Badiner (Berkeley: Parallax Press, 1990), 162.
57. Chao-Fei, *Buddhist Ethics*, trans. by Wan-Li Ho (Taipei: Fa Jie, 1995), 62–63.
58. Chao-Fei, *Tired Birds Still Fly*, 97.

59. Chao-Fei, *Opposing the Powerful*, trans. by Wan-Li Ho (Taipei: Fa Jie, 1994), 12.

60. Chao-Fei, *Tired Birds Still Fly*, 87.

61. Chao-Fei, *Opposing the Powerful*, 22.

62. Chao Fei, *Tired Birds Still Fly*, 75.

63. *Animals' Voice of Taiwan* 22 (2000): 20.

64. Chao-Fei, *Tired Birds Still Fly*, 83, 180.

65. Chao-Fei, *Tired Birds Still Fly*, 186.

66. Idelette van Papendorp, *The Great European Circus in Taiwan: Not So Great, After All*, Report from the Life Conservationist Association (January 1996): 23.

67. Sakya Chao-Fei, interview by author, June 28, 2000.

68. Chao-Fei, *The Testing Ground of Humanistic Buddhism*, trans. by Wan-Li Ho (Taipei: Fa Jie, 1998), 257.

69. LCA press release, June 16, 2000.

70. Chao-Fei, interview by author, June 28, 2000.

71. Chao-Fei, *Tired Birds Still Fly*, 84.

72. Chao-Fei, *Tired Birds Still Fly*, 183.

73. Sakya Chao-Fei, interview by author, June 28, 2000.

74. Chao-Fei, *Tired Birds Still Fly*, 176.

75. Chao-Fei, *Today's Commentary on Monastic Rules*, trans. Wan-Li Ho (Taipei: Fa Jie, 1999), 380.

76. Jiang Tsan-Teng, "The Greatest Buddhist Thinker in Contemporary Taiwan," in *The Search for Human-Realm Pureland—The Study of Modern Chinese Buddhist Thought* (Taipei: Dao Shyng, 1989), 229–34.

77. Janet Biehl, *Finding Our Way: Rethinking Ecofeminist Politics* (Montreal: Black Rose Books, 1991); Mary Mellor, *Feminism and Ecology* (New York: New York University Press, 1997); Rosemary Ruether, ed., *Women Healing Earth* (Maryknoll, N.Y.: Orbis Books, 1996).

78. Janis Birkeland, "Ecofeminism: Linking Theory and Practice," in *Ecofeminism: Women, Animals, Nature*, ed. G. Gaard (Philadelphia: Temple University Press, 1993), 31.

79. Mary Ann Hinsdale, "Ecology, Feminism, and Theology," in *Readings in Ecology and Feminist Theology*, ed. Mary Heather MacKinnon and Moni McIntyre (Kansas City, Mo.: Sheed and Ward, 1995), 201–202.

80. N. Morton, *The Journey Is Home* (Boston: Beacon Press, 1985), 98.

8

The Con-spirando Women's Collective: Globalization from Below?

Mary Judith Ress

Introduction

Mary Judith Ress documents the founding and the continuing path of Con-spirando, an ecofeminist collective and journal in Chile. Con-spirando's work led to a growing consciousness of the relationships between ecology and spirituality for women in Chile and other southern countries. Con-spirando began with a small group of women who learned about ecofeminism and created rituals, workshops, and a vibrant journal. Mary Judith wonderfully documents how consciousness changes and collaborative efforts can succeed.

L IKE MANY OLD-TIME ACTIVIST FEMINISTS, I have become convinced that net-working is a postmodern way to organize—a kind of globalization from below. I also suspect that we are witnessing a silent revolution—a new grass-roots transnationalism appears to be emerging as our networks challenge tra-ditional notions of nation-states and world markets and their power over us.

Networks have been described as those horizontal pathways of communi-cation and exchange that bypass the hierarchical structures of markets as well as the bureaucratic structures of governments.[1] They can be seen as organized efforts at pleading the cause of others or of defending a cause or proposition. They seek to tap the interest of other like-minded folks in order to enlist them to pressure the powers-that-be to change the status quo. Non-governmental organizations, commonly known as NGOs, are without a doubt advocacy net-works par excellence: they weave both formal and informal webs of connec-tions in their efforts to frame the issue in plausible, well-informed language that will invite us to pitch our hat into the ring and "do something about it." These networks manage to embellish cold facts with testimony from real peo-ple, giving concise, accurate information that might even contain a dash of humor. They are able to tap into the energy around issues we feel strongly about and enlist us to take action in a variety of ways and degrees. If you be-long to this growing network of networking, you already know that much of the connecting is happening through e-mail; and while we might decry the fact that our e-mail boxes are about to burst, we cannot deny the fact that these information flows appear to be creating a new kind of global public.

I am a founding member of Con-spirando, a Chilean-based women's col-lective working in the areas of ecofeminism, theology, and spirituality throughout Latin America. Begun in 1991, Con-spirando is both a network and part of a web of ever expanding networks. We are part of the explosion of NGOs that has emerged during the last thirty years, creating a viable, sophis-ticated global force come of age.

Con-spirando began modestly enough. In 1991, several friends and myself began to invite interested women to come together for creative ritual and re-flection from a feminist perspective. Sometimes we were only a handful of women; at other times our numbers reached twenty to thirty. Word spread through friends and friends of friends. We took turns planning the rituals and they naturally reflected the spiritual quests of the coordinators. In retrospect, I see this as a key "foundational moment" for what would become Con-spirando. During these rituals, we would share our stories, our heartaches, and our heart joys through drama, dance, music, and poetry; through earth, fire, water, and wind; through native Mapuche or Aymara chants and drums; through silence; often through tears. Without exception, those of us who form or who have been members of the Con-spirando collective agree that *it was these rituals that brought us into being* and that, in some very fundamental way, define us as a group. The intimate sharing—as well the search for more authentic spiritualities—created a deep bond among us and an enthusiasm to communicate what we were experiencing with other women throughout Latin America.

Although it was feminist ritual that first brought us together, by mid-1991 we were planning to convoke a network that would have three characteristics: it would be marked by a feminist perspective; seek a spirituality and a theology that would be more adequate for women; be committed to the earth as both sacred and as source of life—and therefore share a great anguish at the planet's destruction caused by patriarchy.[2]

Some background is provided to this dream of forming a network by a letter I wrote in July 1991 to women I knew were directly interested and working in feminist theological issues in Latin America. At that time, I wrote:

> I am trying to get a handle on what I'm convinced is a burgeoning grassroots feminist spirituality and theology movement throughout Latin America. My hunch is that, while expressed in many different hues and shapes, Latin American women are thinking about both God and church in new ways and are creating new rituals and spaces to reflect those insights.
>
> I don't know if either the Latin American feminist movement or any church group within the region has tried to take the pulse of women's growing interest in feminist spirituality, but I personally feel the urgency to find out what sort of reflection is going on so we can share our insights and form some kind of network—a web of sisterhood.[3]

The letter included a questionnaire asking for information about existing groups involved in feminist theology and spirituality, as well as soliciting opinions about new images of God, rituals, and interest in establishing a network—through, possibly, a publication—that would exchange information and reflection on feminist spirituality and theology. The answer to my query was most enthusiastic, especially from the following key women: Maryknoll Sister Rosa Dominga Trapasso, co-founder of Talitha Cumi in Lima, Peru; Methodist Pastor Mabel Filippini, director of the Centro de Estudios Cristianos, and Safina Newbery, founder of the feminist collective Urdimbre de Aquehua, both in Buenos Aires, Argentina; Gladys Parentelli, feminist journalist and Catholic Church activist in Caracas, Venezuela; Methodist feminist theologian Janet May, at the faculty of the Universidad Bíblica Latinoamericana in San José, Costa Rica; Cristina Grela, coordinator of Católicas por el Derecho de Decidir in Montevideo, Uruguay; and Ivone Gebara, Catholic ecofeminist theologian, in Recife, Brazil. These women became the mentors— or as I prefer to call them, the "midwives"—of our initial efforts to form a network of women engaged in (eco) feminist theology and spirituality throughout Latin America.

By August, we were planning to launch a pilot issue of a magazine that would convoke the network. At that meeting we decided to name ourselves and our magazine "*Con-spirando*," a play on words meaning "breathing with" instead of "conspiring against." At the time the name seemed a way to communicate our ecological perspective of bringing together many women to "breathe with," to circulate new energies throughout the earth.

We launched the first issue of *Con-spirando: Revista latinoamericana de ecofeminismo, espiritualidad y teología* (*Con-spirando: A Latin American Magazine of Ecofeminism, Spirituality and Theology*) on International Women's Day, March 8, 1992. In our editorial we invited our readership to "self-convoke to form a network of women in Latin America who yearn to have a spirituality and a theology of our own that more faithfully reflects our experiences of the Holy."[4]

Written by team member Elena Aguila, our rallying call published in that first issue of *Con-spirando* continues to define us. In part, that defining purpose read:

In the patriarchal culture in which we live, women's contributions are not taken seriously. This is particularly true in the area of theology. Women are absent as subjects doing theology and also as a major subject matter of this theological reflection. Our lives, our everyday religious practice and our spirituality is simply not present in current theological reflection. Absent too, are our experiences of suffering, joy and solidarity—our experiences of the Sacred. Besides expressing our criticism of patriarchal culture, we also seek to contribute to the creation of a culture that allows theological reflection to flower from our bodies, our spirits—in short, our experiences as women. We seek theologies that take account of the differences of class, race and gender that so mark Latin America. We hope to open new spaces where women can dig deeply into our own life experiences without fear. These experiences are often negative, even traumatic, in terms of the religious formation we have received. We seek spaces where women can experience new ways of being in community; where we can celebrate our faith more authentically and creatively; where we can rediscover and value our roots, our history and our traditions—in short, to engage in an inter-religious dialogue that helps us to recover the essential task of theology, which is to search out and raise the questions of ultimate meaning. We are convinced that, to bring about relationships marked by justice and equality, we must celebrate our differences and work toward a greater pluralism worldwide. To this end, we need theologies that unmask the hierarchies in which we live, theologies that, rather than seeking to mediate Mystery, celebrate and explore the Holy without reductionisms or universalisms. We call for theologies that question anthropocentrism and that promote the transformation of relationships based on dominance of one race, nationality, gender or age group over another and of the human over other forms of life. Such theologies will have profound political consequences. Such a feminist perspective based on our diversity of class, race, age and culture must also take up our love as well as our anguish for all life on the planet, which we feel, is so threatened today. We call this posture ecofeminism. It is within this perspective that we seek a spirituality that will both heal and liberate, that will nourish our Christian tradition as well as take up the long-repressed roots of the native peoples of this continent. We want to explore the liberating dimensions of our experience and imagination of the Holy. To do this, we *"con-spirar juntas."*[5]

In 1992 we published two issues of *Con-spirando*, and since 1993 we have been publishing four issues each year. Each issue treats a specific theme from an ecofeminist theological/spirituality perspective (topics have ranged from sustainable economic systems, to indigenous peoples, death, gender relations, Jesus, healing, embodied education, myths and their power over us, etc.).

In 1993 we were able to rent an old house in downtown Santiago and settle in to carrying forward our overall objective of enabling women to reflect theologically on their experiences of the Holy in order to see them as both legitimate and empowering. We've defined specific objectives: At the Latin American level, we wish to create and nourish a network of women concerned with the themes of ecofeminism, spirituality, and theology, specifically through the publication of a quarterly magazine, *Con-spirando*; in Chile, we aspire to provide a space for women of different economic and religious backgrounds to come together to share our experiences of the Holy through ritual, study, and reflection; and we work to prepare women in feminist theology and ecofeminism through ongoing workshops to empower them to become grassroots feminist theologians.

For the past ten years, then, we have been giving flesh to these objectives through publication of the magazine, through a yearly cycle of feminist rituals that are open to the public, and through a variety of workshops, seminars, and conversations in feminist theology, gender studies, and ecofeminism, including an annual two-week summer school on ecofeminist spirituality and ethics that concentrates on myths and their power over us.

As one small—but resilient—patch in the overall quilt, Con-spirando continues to be recognized as a collective of women, as a magazine, and as a gathering place that seeks to offer an alternative vision of "how things might be" in a postpatriarchal world. I am convinced that our strength lies in the way we have organized ourselves: as a collective we are deeply committed to a circular rather than a hierarchical organization structure. While functioning this way is an ongoing challenge, we all agree that it is a very rich way to work and may indeed serve as a model for others. As a collective we have learned to appreciate each others' talents and skills and have discovered new areas of expertise that we didn't know we had. We've also had to learn to cut through the division between "professional" and "nonprofessional" jobs, because all of us have to do secretarial, cleaning, accounting, and jack-of-all-trades sorts of tasks besides editing a magazine, giving workshops, and coordinating rituals. Through our way of organizing ourselves, we are convinced that we offer an alternative to the current pyramid structuring of most enterprises.

When I review our history, I confess to a belief in the synchronicities of life: Con-spirando happened because the time was ripe, the right people were in place, and there was a real thirst for ecofeminist theology and spirituality in Chile in the early 1990s. However, absolutely pivotal to our sense of who we were and what our direction might be was the visit in March 1993 of Brazilian ecofeminist theologian Ivone Gebara. Invited by the Diego de Medellin Ecumenical Center's yearly program of "Semanas Teológicas de la Mujer," Ivone's course had enormous influence on us. Situating herself within Latin American theology, she described what she called "holistic ecofeminism" as a postmodern challenge to the current anthropology and cosmology of Christianity. Without mincing words, Ivone was able to pinpoint with razor-sharp clarity the patriarchal underpinnings of Christian theology in relation to God, Jesus, sin, and resurrection. She invited us to relativize Christianity by seeing it as only one way to try to understand Holy Mystery. At the same time, she insisted that she was not "post-Christian," but only "post-patriarchal"! We have been fortunate to have her inspiration, and we are also happy that she considers our work effective: "At the moment, in my view, the liveliest ecofeminist group in Latin America is the Con-spirando Collective in Santiago, Chile. With great effort and courage, the collective has been publishing the ecofeminist journal *Con-spirando* since 1992."[6]

I want to emphasize that Latin American feminist theological reflection, including Con-spirando's, is deeply marked by the concrete, lived experiences of women. Influenced by liberation theology's option for the poor, feminist theologians stress the feminization of poverty: "the poor have a face and it is the face of a woman and her children," has become the starting point for much of our theological work. New symbols of the sacred, reinterpretations of biblical texts, challenges to patriarchal church doctrines, a more expanded understanding of ethics, especially in the area of sexuality—all these theological issues are being addressed from the heart-wrenching context of women's tears, suffering, anxiety, and fears as well as our joy and hope.

Having situated Con-spirando in Latin American feminist theology's evolution, what has been our specific contribution? Several of us are uncomfortable with the very word *theology* because it is such a patriarchal term and links us in the popular mind to *religion, church, morality, doctrine,* and, of course, to *the science of God.* Sometimes we think we should just drop the term *theol-*

ogy from our name altogether! At other moments, we renew our commitment to push the definitions of theology past their patriarchal confines by broadening the term to grapple with "constructs of meaning" as concretely experienced by women. We are dedicated to bringing to bear the insights coming from anthropology, psychology, literature, and gender analysis to our understanding of theology.

Although this list is not exhaustive, I believe Con-spirando has made a significant theological contribution in the following four areas: unmasking some aspects of theological violence toward women; renaming and connecting with the sacred; offering an embodied theology; and bringing an ecofeminist perspective to theology. To delve deeply into each of these areas is well beyond the scope of this essay. However, I briefly summarize the work we are doing in each.

Influenced by feminist/womanist theologians such as Dolores Williams, Rita Nakashima Brock, Joanne Carlson Brown, and Rebecca Parker,[7] we have been working toward a nonsacrificial reading of redemption in order to liberate Christianity from patriarchy. With these theologians we have been challenging Christianity's core doctrine that Jesus' death on the cross was essential to redeem humanity from sin. Indeed, helped by their analysis, we are seeing that the theme of atonement and redemption in Christ may be directly related to "allowing" violence and child abuse. According to Rita Nakashima Brock, "the father allows, or even inflicts, the death of his only perfect son. The emphasis is on the goodness and power of the father and the unworthiness and powerlessness of his children, so that the father's punishment is just, and the children are to blame."[8] We have translated the essay "For God So Loved the World" by Joanne Carlson Brown and Rebecca Parker into Spanish and have used it in our workshops. This essay argues that the central image of Christ on the cross as the savior of the world communicates the message that suffering is redemptive—a message that is further complicated by a theology that holds that Christ suffered in obedience to his Father's will. The authors criticize Christianity's central belief of Christ's suffering and dying for us as accepting and even encouraging suffering and ask if it is really so strange that there is so much abuse in society when the predominant theological image of our culture is that of divine child abuse!

In 1994, *Con-spirando* dedicated an entire issue to the underpinnings of violence, paying particular attention to theological violence.[9] In a more recent issue, Bridget Cooke and Ute Seibert grappled with the theme of child abuse and of images of God related to the concept of power. Their reflections grew out of a workshop on images of the sacred in our understanding of power. Participants discovered that the patriarchal image of an all-powerful God who rules over humans had impeded them from understanding that "all of us form part of each other, and we co-create each other in the depths of our being."

Violence and "beyond violence" was the theme for the Shared Garden theological program, organized jointly by Con-spirando, WATER, and Ivone Gebara held in 1997–1998. During each of the three "Gardens," which were held in Santiago, Chile (January 1997), in Washington, D.C. (June 1997), and in Recife, Brazil (July 1998), participants worked with Ivone Gebara on unmasking the myth of Adam and Eve's "fall" from paradise; they also were exposed to Elisabeth Schüssler Fiorenza's concept of *kyriarchy* to describe the system of hierarchical power relationships that impinge upon us at every level. At each stage, women were encouraged to untie the knots of violence embedded in the memory of their individual bodies, as well as in the "bodies" of church, society, and theology. At the same time, they were encouraged to move "beyond violence" to nurture new theologies based on women's experience.[10]

To speak one's own theological word, to name and reflect upon one's own experience of the Holy is essential in the process of elaborating our own theology. Offering new images of the sacred—out of which evolve both new ethical demands as well as new spiritual practices—has been part of Conspirando's work since our beginnings. In our magazine, workshops, and rituals, we try to empower women to rename the sacred according to their own lights. New images that have surfaced include: a pregnant woman giving birth; a great uterus as the body of God; a nest; a tree; a mountain; a flowing river; the ocean; a gentle breeze and a wild wind; a web; a hungry child, an elderly invalid; a circle of laughing children; the sunset—to name a smattering. In workshops we invite women to draw, mold, and dance their sense of the sacred. We often ask them to fill out a chart that shows their personal evolution of images/symbols of the sacred through childhood, adolescence, and adulthood.

In March 1997 we dedicated an entire issue of *Con-spirando* to the creation and evolution of symbols. In the lead article, Josefina Hurtado describes how symbols of the sacred reflect the belief system of an entire people, an entire historical period, and as such contain tremendous power over that people and historical context. However, Josefina stresses that symbols of the sacred are created by we humans in response to our experiences—and therefore they can and should evolve.[11]

That evolution is expressed quite clearly in the following testimony by Yeta Ramirez of Nicaragua:

Ever since I was eight years old, I was surrounded by images of the saints and of the Virgin in the homes of my aunts. I remember one aunt in particular who lived in the country and every afternoon she would sit in front of her altar with all of her statues and pray the rosary and other prayers. We kids used to laugh at her because she was always interrupting her prayers to give orders about domestic matters, etc. At that same time, every Saturday I had to go to catechism classes to receive instructions for making my First Communion. What symbols of the Sacred did I receive then? The image of Christ on the Cross, Christ in his casket that was paraded through the streets every Holy Week. The Sorrowful Virgin was also very present; she was dressed in black and was crying at the foot of the cross. Later on, as an adolescent, I began to work with the Catholic Action movement. At that time, I was very impressed by colors—the white of the Virgin and of the Resurrection. In the 1960 liturgical reform, I began to discover the Risen Jesus who was with us, and the historical Mary, the *campesina* woman who baked bread. Work became a sacred symbol as well as the tools we used for work. We also began to integrate the struggle for liberation into our symbol system. The color green was important—giving a green light for Latin America to take over the land, for agrarian reform. Green was present in our liturgies as a symbol of commitment, hope, and the struggle for land.

Then, in the 1980s, I began to go deeper into nature and incorporate this in liturgy. In my time with Catholic Action, a symbolism of God in nature was present as well as a certain contemplative dimension. More recently, I have taken up an ecological perspective as if I was finally remembering something that I had somehow forgotten. Also, in this past decade, there has been a renewed interest in rediscovering our own cultural roots, which have been repressed. Now we no longer speak of father god, but also of mother-god and of the goddesses.

All this symbolism comes out when I work in the area of ritual and liturgical celebration. In the women's group that I belong to (which is a group committed to stopping violence in all its forms), we've celebrated rituals of healing where we use oils and plants, water, flowers, candles, food, etc. We are now at a stage where we are reclaiming our bodies as sacred in order to combat all the physical and psychological violence we have been subjected to. We have come to see that the concept of "god" is very closed and that really the divinity is much, much broader. We have come to understand that the churches do not have the power to name sacred symbols—that power is in the people and so it is we ourselves who must discover those symbols that speak of the divinity for us. This allows us to build spirituality from our own conception of what is divine. The religious symbols of my childhood were all given to me from outside; now, however, I mold them and give them meaning based on my own experience.[12]

Con-spirando's third theological contribution is in the area of methodology. We espouse an embodied theology holding up women's bodies as "sacred text." Our method has developed out of years of working with the *concientización* methodology developed by Brazilian educator Paolo Freire where oppressed groups, by concentrating on their own experience, engage in social analysis for change (praxis).[13] What we have learned through our work with women, however, is that our bodies are social and cultural constructs, that our history of violence and pain as well as joy and pleasure is stored in our body's memory. The body, then, becomes our theological starting point: to counteract the patriarchal mindset that a woman's body is the source of evil; to heal the dualistic split between body/spirit; and to learn to love our bodies, discovering that we are indeed embodied "temples" of the Holy.

Much of our pastoral work has to do with healing, in workshops where we introduce simple practices such as Tai Chi, deep breathing exercises, and hand, foot, back, head, and shoulders massage techniques. Real transformation takes place. Healing touch, more often than not, awakens new life and participants tell of feeling loved and cared for, as well as safe and protected. We promote learning about and returning to some of the age-old healing practices of our indigenous ancestors who did not separate physical and psychological illnesses because their concept of health was based on a power to heal that comes from within.

We have also discovered the intrinsic connection between our bodies and our spirituality. Since our beginnings we have celebrated the link between wholeness and holiness and have developed rituals where we pray through body movement, dance, and chant. We have learned to believe in this embodied methodology as a more holistic, intuitive way of learning where we become aware of the interconnectedness of all in all. Again, we have dedicated an entire issue of *Con-spirando* to the theme of embodiment as a method for both personal and cultural transformation.[14]

Finally, the fourth contribution we make to theology is our growing commitment to an ecofeminist perspective. While ecofeminism is not native to Latin America, over the years we have come to identify with its intuitions— some of us more enthusiastically than others, and always with our own qualifiers. For Con-spirando, ecofeminism is a new term for an ancient wisdom— a wisdom that still lies dormant deep within our genetic memories. It is the slowly dawning rediscovery that we are not "masters of the universe" but rather form part of the great web of life with all creatures, great and small. Ecofeminism's greatest insight is the (some would say romantic) notion that everything is connected, and therefore everything is sacred. We humans, then, are of one fabric with all life on this glorious planet we call home. Ecofeminists make the connection that the oppression of women and of people of color by a system controlled by ruling class males and the devastation of the planet are not only two forms of violence that reinforce and feed upon each other, but that they both come from a terribly misguided sense of the need to control, to dominate the Other, that which is different (in short, the patriarchal mindset). From being the source of life, both women and the earth have become resources.

As ecofeminists, we join with all those searching for a more holistic worldview that recognizes and celebrates the web of all life. Kinship. There is no Other—the Other is myself. While rejoicing in diversity, which makes that web vital and strong, we know that we are one *com-union*. And it is from that knowledge that we find the energy to engage in action for change, convinced that today there can be neither social justice nor gender equality apart from the well-being of the earth's ecosystems.

As mentioned above, we have been influenced by Ivone Gebara's "holistic ecofeminism." In her 1999 book, *Longing for Running Water: Ecofeminism and Liberation*, she writes:

> With ecofeminism I have begun to see more clearly how much our body—my body and the bodies of my neighbors—are affected not just by unemployment and economic hardship but also by the harmful effects the system of industrial exploitation imposes on them. I have begun to see more clearly how the exclusion of the poor is linked to the destruction of their lands, to the forces that leave them no choice but to move from place to place in a ceaseless exile, to racism, and to the growing militarization of their countries. To defend the unjust monopoly of a minority, the poor countries have become more intensely militarized: they arm themselves to kill their own poor. I have come to see how much all this fits in with the inherent logic of the patriarchal system, especially in its current form. . . . I sense that ecofeminism is born of daily life, of day-to-day sharing among people, garbage in the streets, bad smells, the absence of sewers and safe drinking water, poor nutrition, and inadequate health care. The ecofeminist issue is born of the lack of municipal garbage collection, of the multiplication of rats, cockroaches and mosquitoes, and of the sores on children's skin. . . . This is no new ideology. Rather, it is a different perception of reality that starts right from the unjust system in which we find ourselves and seeks to overcome it in order to bring happiness to everyone and everything.[15]

The Con-spirando collective attempts to offer an ecofeminist perspective to every theme we cover in the *Revista Con-spirando*. We have also dedicated three entire issues specifically to ecofeminism.[16] We give workshops on ecofeminism and ecofeminist theology and spirituality, which have been and continue to be our strength. The Shared Garden program and our School of Ecofeminist Spirituality and Ethics also have a strong ecofeminist component in content and methodology. Our rituals, in a zillion creative forms, convey that we are one Sacred Body, with all its welcome diversity! Our ecofeminist posture places us firmly in the postmodern debate as well as in the post-patriarchal (but emphatically *not* "post-Christian") quest for a more relevant and passionate understanding of those with whom we are in relation. As it turns out, we may well be in relation to the entire cosmos, according to recent scientific discoveries. This posture has engaged us in the struggle to redefine divinity as well as the human enterprise. We search for a more adequate cosmology, ethics, and spirituality. And it is this search—rather than the tentative answers—that bind us together as a collective.

Over the years, we have come to see that our ability to influence the status quo is in advocating for cultural over political change. (This appears to be a general trend here in Latin America: those of us who were leftist militants in the late 1960s and early 1970s, then staunch resisters to the military dictatorship in the late 1970s and 1980s, now find ourselves worn out from our past militancy and wondering what was achieved in the long run.) We also find ourselves less attracted to massive mobilizations and protests, preferring small groups and local issues as our locus for weaving new visions and nurturing energies to edge those visions forward.

Concretely, this has evolved into a commitment to sustainability, to embracing local initiatives that enhance the bioregion. Both individually and as a collective, we try to live alternative lifestyles: we compost, recycle, try to eat organically, seasonally, and lower on the food chain; some of us eat our own homegrown veggies and all of us have quit smoking! We belong to a fledgling initiative that is trying to reintroduce barter and local currencies to buy locally produced goods; we give plants for birthday gifts and plant trees in memory of loved ones who have passed on. We belong to RENACE, Chile's feisty environmental movement, and we actively support a growing call for a national economic plan that would allow Chile to develop a "steady state" economy instead of being driven by GNP indicators that neglect to register the cost of the country's so-called economic miracle. This purported "miracle" has led to the destruction of Chile's old growth forests, the overfishing of its coastal waters, lakes, and rivers, the increasing desertification of the north, the huge imbalances and pollution caused by the large-scale mining of copper, the industrial smog that covers Santiago, making it one of the most toxic capitals in the world—the list goes on and on.

We know that we are living in "in-between" times, with the utopian visions of yesteryear now only a nostalgic memory, and the ecofeminist dream still to be made flesh. Yet we see ourselves as one of the small "communities of celebration and resistance" described by ecofeminist theologian Rosemary Radford Ruether: while trying to unplug as much as possible from the "principalities and powers" of the present system of late capitalism, we try to live, work and celebrate the Holy in our lives in a way that invites a more holistic way of living gently on this planet, Home.[17]

What are possible clues for transformation that the Con-spirando experience might offer other women's groups in Latin America—and beyond? What new ways of acting or new insights might we offer? First, there is no doubt that as we open a new century (and millennium), ecofeminist theology and spirituality is becoming a dynamic passion and posture for many grassroots women's groups in Latin America. (However, the great majority of these groups are emerging from women of faith—religious sisters, pastors, pastoral agents, leaders/members of Christian Base Communities [CEBs]). When Con-spirando issued its rallying call in 1992, we were the only collective dedicated to pursuing an ecofeminist theology and spirituality. Today, besides the already existing collectives that have assumed and deepened an ecofeminist commitment (Talitha Cumi in Lima, Peru; NETMAL in Sao Paulo, Brazil), new ecofeminist reflection groups or collectives have been formed in Bolivia (Santa Cruz and La Paz), Uruguay, Venezuela, Mexico, Honduras, El Salvador, Guatemala, Ecuador, and Chile (Concepción). How strong these groups are, and their impact on their local surroundings, remains to be researched. But without a doubt the network is growing.

Probably more than any other factor, it is the passionate eloquence and commitment of Ivone Gebara that has ignited and fanned the fire of these groups. Calling herself a "nomad," Ivone travels to the farthest reaches of Latin America to "accompany" groups (mostly women's groups) seeking a more holistic theology and spirituality. It is not simply her evocative ecofeminist perspective that attracts (mostly poor) women to her courses; it is her sisterly concern for each, her compassion and solidarity that she communicates through a look, a gesture, and a hug. Yet, by her own admission, Ivone is one who adds the *salsa a la sopa* (spice to the soup). It has been Con-spirando and our sister groups that have been able to provide the infrastructure for her "spice." Thus, we see ourselves as a key "hinge" in the ongoing development of an ecofeminist theology and spirituality network in Latin America.

Although we are very proud of our accomplishments, we know that ecofeminist thought is only beginning to have an impact on theology and feminist theory in the region. Mainline patriarchal theologians, as well as liberation theologians, are threatened by ecofeminism because of its challenge to orthodox definitions of God and the human in relation to the divine. Feminists criticize ecofeminism as being too "eco" and not enough "feminist." They also warn against identifying women with nature, which would be falling into an essentialist trap that, in the end, would unwittingly support patriarchal dualistic constructs.

For us, a major transformational tool is our commitment to embodied learning. Body prayer, ritual, intuition, and healing practices all offer new ways of learning—not only for women, but for all humanity. And with our yearly cycle of rituals, we give flesh to our commitment to empower women to celebrate the sacred as we see fit. Another tool is the *Revista Con-spirando*, which attempts to keep a pulse on, as well as nourish, ecofeminist theology and spirituality in Latin America. We are convinced that we provide, for a Latin American readership, key information, analysis, and reflection that is not found elsewhere.

There are, of course, voices that caution us to water our Latin American ecofeminism with the streams of real life—the lives of the region's poor majorities. We must beware of essentialism, that is, depicting women as closer to nature because of our cycles, our relationship with the moon and the tides, and so on. We must avoid the label of being "New Age" with an individualist, "make me feel good" spirituality. And we must respond with visible actions to any accusation that an ecofeminist posture avoids justice and human rights issues. We are also very fragile in terms of financial and human resources.

However, a guiding metaphor of ecofeminism is the seed. Instead of the predominant "power over," the seed suggests "power within." The seed lies dormant; it breaks open, sprouts forth, blossoms, bears fruit, matures, withers, and falls to the ground again. It will be what it is meant to be. We too are seeds, called to be what we are meant to be. And so is Con-spirando. May it be what it is meant to be—nothing more, nothing less.

Notes

1. Margaret E. Keck and Kathryn Sikkink, *Activists beyond Borders* (Ithaca and London: Cornell University Press, 1998), 1–4.

2. *Con-spirando*, "Reunión del 'red'" (*Con-spirando* archives, 18 agosto 1991), 1.

3. Mary Judith Ress, letter, 10 July 1991.

4. Elena Aguila, "Editorial," *Con-spirando: Revista latinoamericana de ecofeminismo, espiritualidad y teología* 1 (Marzo 1992): 1.

5. Aguila, "Editorial," 2–5.

6. Ivone Gebara, "Intuiciones Ecofeministas: Ensayo para repensar el conocimiento y la religión," in *Doble Clic* (Montevideo, Uruguay, 1998): 38.

7. Dolores Williams, *Sisters in the Wilderness: The Challenge of Womanist God-Talk* (New York: Orbis Books, 1993); Rita Nakashima Brock, *Journeys by Heart: A Christology of Erotic Power* (New York: Crossroad, 1991); Joanne Carlson Brown and Rebecca Parker, "For God So Loved the World?" in *Christianity, Patriarchy and Abuse*, ed. J. Carlson Brown and C. Bohn (Cleveland: Pilgrim Press, 1989).

8. Brock, *Journeys by Heart*, 56.

9. *Revista Con-spirando* 8 (Junio 1994).

10. Con-spirando Collective, "Sistematización: Más allá de la violencia: solidaridad y ecofeminismo" (Summary of Shared Garden held in Santiago, Chile, January 27–February 8, 1998. Manuscript prepared by Con-spirando).

11. Josefina Hurtado, "Por sus símbolos los conoceréis," *Revista Con-spirando* 19 (Marzo 1997): 2–9.

12. Yeta Ramírez, "El poder de los símbolos," *Revista Con-spirando* 19 (Marzo 1997): 35.

13. Paolo Freire, *Pedagogy of the Oppressed* (New York: Herder and Herder, 1970).

14. *Revista Con-spirando* 26 (Diciembre 1998).

15. Ivone Gebara, *Longing for Running Water: Ecofeminism and Liberation* (Minneapolis: Fortress Press, 1999), 2.

16. "El ecofeminismo: reciclando nuestras energías de cambio," *Revista Con-spirando* 4 (Junio 1993); "Etica y ecofeminismo," *Revista Con-spirando* 17 (Septiembre 1997); and "Ecofeminismo: hallazgos, preguntas, provocaciones," *Revista Con-spirando* 23 (Marzo 1998).

17. Rosemary Radford Ruether, *Gaia and God: An Ecofeminist Theology of Earth Healing* (San Francisco: Harper, 1992), 268–74.

9

Ecofeminism: An Ethics of Life

Ivone Gebara

Introduction

Ivone Gebara, an ecofeminist theologian, opens up the connections between social ethics and ecofeminism in the context of Brazil. Ivone seeks relationships between feminist and ecological thought in everyday practice. Under the rubric of an "ethics of life" she seeks answers to the "great questions" of war, racism, and other violence. Ivone takes the historical and Christian idea of women as nonethical subjects who are close to a debased nature and shows how this has been lived and understood in Brazil. Ivone reinforces the need for ecofeminism to be cognizant of history and ethics and asserts that any discussion needs to be contextualized.

From Ivone Gebara, "Ecofeminism: An Ethic of Life," in *Sacred Earth, Sacred Community, Jubilee, Ecology, and Aboriginal People* (Toronto: The Canadian Initiative, 2000), 29–46. Copyright © 2000 by the Canadian Initiative. Reprinted by permission.

I WOULD LIKE TO BRING A CONSIDERATION of social ethics into my reflection on ecofeminism; that is, a consideration of effects on our actual behavior. I hope to demonstrate the opening and articulation of feminist thought to ecological thought in our everyday practice. My perspective is based on the need to seek paths toward answers to the great questions of our time: the proliferation of wars, the displacement of entire peoples, racism, political and religious dogmatism, the exclusion of people from work, and many other forms of violence. The latter represents the path of capitalism in its current form of internationalization of capital for the benefit of a privileged minority. I firmly believe that ecological issues and the growing exclusion of entire peoples are linked to the global exploitation and the spiral of violence that characterize our world.

It is important here not only to consider economic issues, which are obviously paramount, but in particular to criticize the limiting hierarchical and Darwinian view of human beings in which humans are seen only as those who dominate other humans and the Earth. This understanding of what is human is based on an ethical "claim" and way of acting toward other beings with respect to earthly goods and the balance of ecosystems. It follows from a hierarchical view of humans based on profit and a quasi-definition of humanness based on the capacity to own, to know, and to have power. In this view, certain men claim and enjoy the prerogative of being called "human" under the current social system. I feel compelled to try to change this logic; those who feed on the blood of others, who benefit from the destruction of forests, and who produce chemical garbage should really be called "subhumans," viewed from the perspective of ethics and reciprocity.

One of our greatest challenges at the dawn of the third millennium is the urgent need to reappropriate our sense of ethics. But what does ethics really mean? Who defines it? Where does it come from? Do women have a specific ethics? How should we conceive of ethics based both on concrete contexts and in relation to our global context? How can we think of ethics beyond the universality and idealization of concepts that belong to patriarchal systems? All these questions provide a background for this reflection, which I see comprising three main areas: Women as Ethical Nonsubjects; Women and Nature: Relationships and Conflicts; A Prelude to an Ecofeminist Ethics.

The hierarchical, patriarchal, classist, and sexist world in which we live teaches us that certain privileged people determine all proper behavior for us. We normally expect authorities to fulfill this role, as if they were somehow endowed with a higher morality. Under this belief, certain groups, particularly women, have historically not been considered as having equal abilities, rights, or even citizenship. Of even more concern is that such persons have not even been considered completely ethical beings. That is, they are not invited to take part in private or public decision making on issues that affect them. It is also unfortunate that these people themselves often believe that they cannot be historic subjects in the full sense of the term. Their consciousness is more or less asleep, and in the countries of the "Two-Thirds World," the struggle for survival makes it even harder to awaken consciousness. To be more precise, the consciousness of poor women is tuned in to their personal situation, but these women are prisoners of their daily lives and have no access to the power needed to make effective change in their lives. It is as if the key to their prison were kept in faraway lands; the women do not have the strength either to knock down the bars or to get new keys made.

The situation of women in the "Two-Thirds World" has undoubtedly improved through the efforts of various human rights and women's rights movements, as well as through the effective actions of many feminist movements. However, despite this progress, it is still normal for women to live as ethical nonsubjects. This belief and way of operating are still the rule in our relationships at all levels of society. We submit to laws that negatively affect women as if they represented pre-established values and, to some extent, standards imposed on us to guide our behavior. Who says that our work is less important than men's and that therefore we should be paid less? Who determines what social and medical services we need? Who establishes the division of social labor and responsibility? Who defines sin and guilt for us?

If this situation holds true in civil and economic society, it applies equally to religious institutions, especially Christian ones. Who determines the images of God in our societies? Who defines the relationship we should have with God or Jesus Christ? Who decides on women's sin and proposed salvation? Who determines their authority, or lack of the same, within churches? It must be recognized that women in our churches and theology have been treated as ethical nonsubjects, incapable of making important decisions in their lives or the lives of their communities. Women's religious experience and way of perceiving the mystery of life have never been taken seriously by religious or clerical institutions dominated by male power.

How can we understand this situation and make a change toward fairer and more supportive relationships? There are different points of entry that can help us arrive at a view of this problem. I open the door to one of them here, in the realization that it alone is not enough to address the problem in all its complexity. Others are needed as well. I am trying to open the door of Christian morality a crack, specifically the morality of Roman Catholicism as it is lived in Latin America. I rely on my own observations and life experiences as well as those of other women, who, like me, are seeking new ways to live justice and solidarity.

In a general way, we know that the precepts of Catholic morality have been used in the form of laws for confessors. The training of confessor priests has been of great concern for the Church, since it is through their agency that the control (called salvation) of bodies and souls is assured. The accent has been on isolated actions rather than on the life of each individual and her or his context. Personal responsibility was measured by pre-established laws rather than by the actual circumstances of a person's life. This tradition strongly emphasizes the identification of canonic law with morality, or Christian ethics, similar to the way today's civil society often identifies morality with written laws. Remembering this helps us to understand our present era, although the ethical issues within churches are discussed today in a different way. We know that legislators are always men and benevolent fathers; they are pastors careful not to lose their sheep, making sure they do not stray from the flock of the children of God. In general, these legislators do not speak on our behalf; they do not consult us, but rather decide and legislate for us, certain as they are that they know God's wishes for us. The limits to this "benevolence" have very specific significance for women.

The social encyclicals since the nineteenth century, for example, the *Rerum Novarum* (1891) and later, the Vatican II Council Papers (1964), moved from an individual morality to a social morality and a social ethics. The concerns of workers, Third World countries, and the social consequences of development for a time displaced individual morality and put the focus on the social axis. There was significant development of social morality, underlining the contradictions of models of development. The same movement took root in Latin America. We developed a major discourse on the ethics of liberation based on conditions of poverty and social injustice. Some bishops, particularly the Brazilian National Conference of Bishops, strongly protested the abuse of political and economic power and its effect on marginalized people.

However, in the last fifteen years, the Catholic ethic has put the contradictions and social injustices aside, especially issues relating to the family, sexuality, and contraception. The ecclesiastical circle, which had opened itself to greater world issues, has closed down on issues of special significance to the lives of individuals. Although it would be possible to look for and reflect more seriously about the cause of this shift, this is not our purpose here. It is nonetheless of interest to remember that the position of the Catholic hierarchy was strengthened as a result of the development of feminist movements and women's growing awareness of their desire to become actors in their lives and active decision makers on issues of concern to them.

We know, even though it is rarely given much attention, that women were the first targets of the writing of the Magisterium relating to the family and family planning. It was our privilege, now eleven years ago, for a pontifical letter to be directed at women: *Mulieris Dignitatem*. However, women's voices, their tears, and their crying out are still unanswered, or more correctly, are still not faithfully reflected. We are talked to; guidelines for living are imposed on us as if these guidelines were laws, of more importance than our own history. We are talked to as if we were children in need of the advice of parents and grandparents. In this perspective, women are once again absent subjects, non-subjects, and noncitizens with respect to the guidelines and decisions of the institutional Church. We continue to be the object of stone-throwing when our behavior deviates from their laws or expectations. At such times, we are reminded of our responsibility as women, that is, our responsibility to be mothers and to defend life against all threats. There is frequent appeal to our nature and life-giving capacity.

This is one of the contradictions present in the Roman Catholic system of ethics, a contradiction marked by a belief in the superiority of men and their role that arises from this sense of superiority in legislating the behavior of women and their bodies. In Catholic morality, evil and sin are defined and legislated in advance. While this is in part understandable, given that we are talking about laws, such identification leads to a reduction of the ethics of law. Ethics is ruled by the world of values, rather than by laws that have been set down. The restriction of legal ethics, far from helping us to develop a sense of collective responsibility for social problems, actually serves to emphasize legalistic, and often static, attitudes. It produces guilty persons and victims, rather than citizens who seek to assume risks and responsibilities in history. In this scheme, women often are guilty parties and victims at the same time. That is, they develop a sense of guilt that is sometimes excessive, coupled with an equally excessive victimization, which seem to feed each other.

The idea that men create ethics and write moral codes is not just a throwback to the Middle Ages but also reflects modern times. Modern rationalism is actually masculine rationalism and is alive and well among us.

Feminist thought about our nonsubject reality requires us to shed our role as guilty parties and victims and take on the role of historical subjects who take responsibility for their actions. To be a historical subject means leaving behind the naïveté in which we have often been held and that we have so well cultivated, to address every situation and the attending risk of decision making, knowing the decisions may be good or bad. Life is a risk; we know that. Now women must take a reasonable chance on themselves, which involves risks for others as well, not just at the level of discourse but in facing our everyday life.

To be a historical subject is to pull oneself out of the ethical dualism patriarchy engenders. For the patriarchy, ethics is founded on the opposing concepts of good and evil. Something is either good or bad, as if such a total separation of good and evil could exist. We, as Christians, have chosen the side of good, as if the good side were obvious and as if Christians were incarnations of good. It becomes increasingly clear that this dualist model for understanding life does not match our concrete, daily experience of life; that is, the daily experience of regular people who wake each morning and must think up a way to live or survive in a society offering few chances for dignified development.

The goodness of women was pre-defined as synonymous with humility, silence, self-effacement, tenderness, and openness. Women are good when they keep quiet and go about their work according to the roles established by patriarchal society. This is why we are afraid to make mistakes and act badly. We are also afraid we will be judged by this established morality. In the end, we are afraid to become human beings who risk their lives in the midst of all others' lives; people who have to learn that life is full of attempts, victories, setbacks, death, as well as experiences of freedom and love. An anthropology characterized by *mélange* is the only path we can follow if we are to understand our humanity in another way.

Various feminist theologians and social scientists have critiqued ecofeminism's rapprochement between women and nature. I quite agree with the criticism of certain essentialist positions that hold the feminine essence as a fact of nature. Such positions accentuate dualism, in which women are seen as closer to nature than men, the primary producers of culture. However, I also believe that the ideological aspect of such a rapprochement is objectionable, and that the ecological struggle is properly the responsibility of both men and women. On the other hand, we must open up our overly anthropological thinking to include consideration of our natural environment. We must stop thinking that human beings are not part of nature, or that culture exists in opposition to nature. We must stop seeing ourselves as extranatural or supernatural creatures who strive to overcome our own human nature and continue our domination over other species.

The world today, characterized by a profit motive without a community ethic and by a lack of concern for marginalized persons and their natural environment, is on the verge of self-destruction. Our governments increasingly invest in war and defense by arms production, especially the production of nuclear and biological weapons. They have forgotten the most basic of all human undertakings: to nourish, protect, and educate a country's citizens. This is considered women's work and therefore represents unimportant and uninteresting policy issues for governments who run the world. Of course, this is not recognized in the public or official record, but is confirmed by a brief look at the budgets of most governments in the world, especially those of the "Two-Thirds World."

Accordingly, certain clarifications are needed, especially with respect to the domination of women and the unjust exploitation of nature for the benefit of a minority. I would like to address these points at two levels: the socioeconomic level and the cultural symbolic level.

We know that within a perspective that embraces the interdependence of all things, all forms of oppression are also interdependent. One oppression can give rise to another or reinforce an existing one. This is true across groups of people, even when one group or another does not experience a particular type of oppression. For example, Black women experience acute skin-color oppression by whites. This oppression is not experienced by white women in the same way. In contrast, sexist oppression is not as marked among primitive indigenous groups as among mixed-race groups in Latin America. The oppressions are therefore not the same and are not experienced by all groups with the same intensity. However, they are closely linked and interdependent, especially in the current context.

We can state without fear of contradiction that the domination of women and nature accelerated under colonization, as part of its political ideology. Therefore, socioeconomic considerations cannot be divorced from the level of cultural symbolism. When Black women were exploited on large sugar plantations in northern Brazil, not only for their farm labor, but also for domestic and sex work, they became a prized sex symbol. Mulatto women are known for the sexual appetite they inspire, their sensuality, their sense of rhythm, and their physical strength. The word "*mulatto*" comes from "mule," a beast of burden that does not complain when beaten by its master. The animal side of nonhuman "nature" is thus affirmed in the body of Black women. Further, the symbolism is closely linked to the reality of economic and social exploitation. This symbolism has entered the music and literature of Brazil, making the country an international success known for having the most beautiful mulatto women.

However, we must distinguish the economic aspect from the symbolic to expose certain elements of the complex issues they might otherwise mask. Note that distinguishing them does not entail understanding them separately.

A discussion of relationships and conflicts between women and nature requires a historical approach. This statement is only valid when situated in the concrete reality of different cultural contexts. I would like to locate myself in the Latin American context with an emphasis on the colonization of the land and its inhabitants. I would also like to underline the colonization and slavery to which African peoples were subjected. Huge numbers of slaves arrived in the Americas during the sixteenth and seventeenth centuries, particularly destined for Portuguese and Spanish colonies.

If the land and the primitive populations of America were occupied, this is no less so for African land and populations. The two processes, colonization and slavery, were joined, giving birth to a much more complex conflict that continues to this day. We know that the women and men were treated differently in the colonial period. This differential treatment generated observable problems in the behavior of men and women. In most countries affected by colonization, there is both a sexual and work-related "colonization" of women's bodies. This is not a matter of essence; it is a historical, cultural, economic, political, and religious matter. The specificity of so-called gender policy, economic gender, and symbolic gender must be taken into account.

Colonization is the occupation of others, through the dimensions of time and space, and the reduction of the identity of the colonized to that of the colonizer, whether this means requiring the colonized to live according to the commandments of the declared colonizer, king, prince, or conqueror. The worst part of colonization is the loss of awareness of being colonized and no longer knowing one's roots, or who he or she was or is. The worst part of colonization is losing self-confidence and one's cultural values, placing oneself in the hands of the other in a submissive and uncritical manner. It is even worse to forget one is colonized and accept things as they are as fate or the nature of life as predetermined by a mysterious and divine will.

The important role of Christian ideology and obedience to the process of colonization cannot be overlooked here. From the perspective of sexuality and work, women's bodies are producers, both biologically and culturally, to the extent that women give birth to and feed their newborns. They do not require any tools to produce, as men's bodies do. However, this productive body is not recognized as an agent of social production. Domestic work, such as rearing and feeding children, is not recognized as work in the social sense, or work basic to the maintenance of life in society. On the contrary, these activities are seen as physiological functions comparable to those of other mammals. It is therefore patriarchal society with its male division of social labor that becomes biology-based and essentialist. Society deprives women of recognition for their daily work and educational and social contributions, which underpin all economic transformational work.

We know that even today, in most countries of the world, breastfeeding mothers, although permitted leave from institutional work, are not fully paid. Capitalism does not consider the role of procreation and feeding as human labor to be valued and evaluated, but as a purely physiological, nonproductive, and noncompensable function of no social value.

The economic and social colonization of women takes various forms as part of colonization. It is not a homogeneous process. There are struggles here and there for decolonization. There have always been some cases everywhere of people becoming aware and working for freedom, attempting to appropriate their own bodies and creativity. This means that women have not failed to react or accepted their fate only passively. Movements and concerted actions have been noted throughout the history of women's resistance to masculine domination. However, we are only beginning to recognize these battles, given our "derivative status," to use Simone de Beauvoir's term. This level involves our production of culture: history, art, literature, sciences, religions, and so on. If we tour the historic monuments in our major cities, with rare exceptions, we will find that history, told through the art of monuments, is the history of "great" men. Battles, generals, and horsemen seem to be the most important figures. However, in representations of justice or liberty, feminine figures are shown.

In patriarchal society, masculine symbolism is dominant even when representing the female body. The body is depicted emphasizing the most desirable or detestable features, from the male perspective. This situation has long been understood as a mark of masculine superiority in the production of culture. Beginning in the second half of the twentieth century, there is an observable awakening in women's historical awareness. We realize that our absence from the production of the symbols of our culture is not due to any lack of natural ability, but to an inappropriate distribution of work, and a relegation of women to domestic responsibilities. This poor social organization of work does not leave women enough time and space to develop their creativity outside of childcare and the home. This is also due to sexist anthropology that makes women second-class citizens. Those who have defied this fate have had to struggle against many prejudices and face high levels of personal guilt, especially those who are mothers.

This is why I hold that women are colonies. That is, women have been colonized to stay home and allow men the grand flights of fancy required for the production of culture, politics, and religion. We assist in the reproduction of a culture that makes men feel they are at the center of history, and its organizers.

Religion has been the justification and the mainstay of the absolute character of masculinity. Christian theology, therefore, has imposed powerful ideological colonization on women's bodies in the name of values said to come from God. The vocation, feminine values, and behaviors of a virtuous woman have been predetermined. Furthermore, theology refers easily to women's "nature" as the "will" of God for humans. If we change what is considered natural for women, we are doomed, because the natural is considered divine. But who establishes what is of God and what is natural? The point is not to emphasize the victimization of women; that would lead nowhere. Rather, we focus on the importance of the awakening of social consciousness, in order to attempt new human relations and new production in all realms of human knowledge. We must understand the rationality and irrationality present in our societies and cultural institutions in order to fight against those that marginalize and oppress women.

We know, as stated above, that the cultural aspect most marked by masculine symbolism is religious, especially Christian, symbolism. Our twenty centuries of patriarchal religious production form a historical and social reference point for a large proportion of humanity. One need only observe to what extent the mystery of the Incarnation, Passion, and Resurrection of Jesus has affected the art, music, and time and space of different peoples who lay claim to a Christian identity. The effect is primarily male. The feminine aspect, when it is present, is often subordinated to the glorification of the masculine. Would there not be a way to create a more inclusive tradition? Would there not be a way to bring a different symbolism into Christianity, one that emphasizes the values women live, fully recognizing these values?

We do not have the wherewithal to answer this question, but I suspect that it would not be difficult to find answers, especially in Latin America. The difficulties are not only linked to issues peripheral to the organization of Christian life, but to the structure of Christianity itself. This structure identifies the divine mystery, which supports and nourishes all life, in the image of one historic man, Jesus of Nazareth. To conceive of a Christian approach that would be more inclusive and respectful of women and nature would require a radical change in the structures underpinning Christianity; these are structures that have philosophically dualist and metaphysical bases. Philosophical deconstruction of Christianity has already been undertaken by various philosophical schools and by radical feminist theology. But this deconstruction has no place within existing religious institutions, as it announces the end of their reign. It also rattles the political and cultural institutions that are based on the same philosophy.

In Latin America today, there is a new Christianity that has been developed by the media. Less moralistic, at least publicly, it is celebratory and aimed at healing immediate ills. It appeals to people's feelings and emotions, and to the various fears that plague people in an increasingly unstable world. This is a Catholicism that calls on symbols, prayers, and traditional religious practices, incorporating a distinctly patriarchal content, and intended to provide people with a sense of security. It gives one a feeling of going back in time, of touching our grandparents' past, a past in which it seemed that God was more attentive to human suffering. Priests and pastors are the mediators in this new dialogue between God and humanity. Attendance at Sunday mass is on the rise and the popularity of the Roman Catholic Church is increasing. Women, unfortunately, are in the majority of those attracted by these movements.

All of the above underlines the immense difficulty of creating new symbols with new content inside Christian communities. We have the impression that a "new colonization" of a cultural nature, often called the "new evangelization," is occurring in countries around the world. Feminism does not appear to have enough strength to take on the complexity of this religious world, although it is putting up courageous resistance.

When we refer to a prelude, we mean the start of something, that is, the beginning of a path that will probably be followed. We mean the announcement of something new in the midst of paths already known. This may apply to a range of situations and human activities. But, in the field of ethics, it becomes very complex to speak of a prelude. In fact, our actions in favor of justice and equality or solidarity are part of a process in which the first of these actions does not guarantee that those that follow will continue in the same direction. To this extent, the prelude to an ecofeminist ethics is undoubtedly the beginning of change marked by new, socially observable changes in behavior. However, it is nothing more than a start to be found among various groups of women and in different places, especially in Latin America.

Therefore, talking about a prelude to an ecofeminist ethics maintains a sense of the temporary. It refers to certain concrete actions and awakenings that announce something new. These fragile announcements carry the mark of vulnerability and human ambivalence. Our personal and social lives are a process that goes through highs and lows, progress and setbacks, that is, the *mélange* that characterizes our being. This means that no ethical behavior today can guarantee the future, and no ethical behavior contains a "promise" of continuity or attainment. We are far from the certainties of modernity. This applies as much at the individual as it does at the collective level. Therefore, the ethical prelude has nothing to do with the consistency we would ideally desire. When a social movement has a moment of success followed by a failure, we seek the cause of the downfall, as if social movements were unaffected by the presence of death. We have not sufficiently integrated the presence of death into social and religious systems.

Ethics is fundamentally a matter for the present, a sensitivity to the now and a commitment made, an action taken today. Ethical behaviors are not necessarily the privilege of elites; they can also occur in those who are not recognized as ethical subjects. In contrast, nonethical behavior can come from people and groups who recognize social and ethical matters. Thus, ethics cannot be identified with pre-established laws, as noted in the first part of this discussion. The laws that maintain their importance and place have a different life span and function. Ethical behavior, beyond pre-established or, for instance, coercive laws, are driven by personal needs, a given moment in individual histories, context, unforeseen circumstances, and suffering. In ethics, there is a desire for freedom, and to some extent, a tendency toward anarchy within us. The wish to go beyond established laws, or even to go against the grain of our own wishes for comfort or profit, may be contradictory.

Transgressing these laws to live ethically may appear just as contradictory, but it is the expression of the experience of many individuals and groups, which explains the difficulty many popular and women's movements experience in trying to agree on concrete ethical positions. An ethics of speech, that is, the protest against injustices, and even the ethics of consciousness, does not necessarily lead to ethical actions involving changed behavior and changes in social institutions. Speaking out publicly and changing one's consciousness represent positive change. Of course, such actions are not enough to change social behavior. The power of alternative words does not have the same force as the word backed up by weapons of war or concrete actions of resistance!

The ethical question is fundamentally that of an ethical life, a life organized around the attempt to take account of the diversity of human groups and of our ecosystem. Ethics begins when the established order, which has been recognized as just, becomes intolerable because it has become unjust. The patriarchal order forfeits its recognition at the point at which the justice it has extolled is unmasked for its injustice in relation to the actual life of women, and is revealed as having a murderous effect on the ecosystem.

The foregoing is linked to the dynamic and changeable character of human relations and the building of sense in our life. This is the reason our organizations working for good always have less good effects and sometimes have bad ones. This is the predictably unpredictable nature of human actions and is why any ethics is always an ethical prelude that can never become a totalitarian and static system. It must always be attentive to the complexity of situations and to the new elements that occur. The ethical within us (or our ethical behavior) certainly leads to good or just actions, but this goodness is also characterized by the provisional nature of life, and therefore becomes an invitation to pursue new ethical steps.

Finally, it should be remembered that the issue of ethics or morality, described in a certain Christian tradition as the balance between self-love and love for others, is no longer a given. Our world shows us that it is truly individual interest, the struggle to dominate others, that seems to be our most "natural" inclination, especially in social practice. Once again, ethics appears as a prelude, an announcement, or a wish for good news that must be given a chance to expand. It is not something to be analyzed outside of the different *mélanges* of human life. It is like a utopian horizon inviting us to look at the consequences of our choices and the structures we have built over the course of human history. It is the utopian horizon, born out of situations at our margins that touch our sensibilities and emotions more. Our yearning for justice, peace, and love is reawakened when these are most lacking. In the situations that stretch us the most—pain, suffering, and catastrophes—analytic reasoning is certainly less powerful.

The novelties of human history have often occurred because of minorities. Feminism is no exception. And in its novelties, feminism is also no exception to the rule of falling back onto the very models it criticizes.

I would like to list some of the ethical perspectives arising in various feminist and ecofeminist movements aimed at changing behavior in human relationships. These include some announcements, and some good preludes beginning to make themselves heard. They allow us to dare to dream of a new moment, even if the moment is wrapped up in all the disorder and oppression of the current world. These perspectives and announcements constitute both gains and recommendations gathered from different activist groups of women around the world, especially from Latin America. They arise from concrete experiences and multiple sufferings and humiliations. I relate them as attempts, desires, and wishes to save them from any claim of ethical systematization that would restrict their mobility and openness to life's various challenges.

1. We use various methods to gain the right to be ethical subjects, to think up our own actions, choices, and relations in the public sphere and to take responsibility for their consequences.
2. We want to achieve the right to make choices about our bodies and end this millennium-old colonization. This is not an egotistical attitude, but one of dialogue and solidarity with others.
3. We emphasize the contextual nature of ethical decisions, beyond the pre-established universals. This does not entail relativism; it is just that universals that erase differences in the lived realities of real people are no longer efficient and do not help us in our everyday lives.
4. We emphasize the importance of concrete events experienced by groups of people, especially marginalized groups and victims of injustices. We seek concrete solutions to their difficulties and the development of solidarity among groups.
5. We insist on the relationality and interdependence of all things, all situations, and all events of human history and the environment in which we live.
6. We wish to emphasize the fact that any ethical position or behavior includes our personal responsibility for good and evil at the same time. Even though I seek good, that good is always part of the *mélange* of life. We must therefore help each other to emerge from our state of innocence and embark on the risks of human history, with its beauty and suffering.
7. We, as women, are no longer, as positivism called on us to be, the upholders of all morality, as though our entire morality rested on our

nonentrance into public history. To enter into public history does not mean to dirty our hands. Our hands are simultaneously dirty and clean in the domestic realm, even if we are not able to assume direct responsibility for social choices we did not make. We must become more involved and dare to build new human relationships, even if this means risking failure.

These and many other leads are part of what I call an ethics of life. It is an ethics that exists within the diversity of the paths life offers us. An ethics of life is an attempt at collective thought and action aimed at creating a society in which every person and species of animal and vegetable have the right to life within a collectivity because each one has a vital need for the other. To this extent, we can say that Christian experience is invited to join in on this ethics of life. This is an ethics undoubtedly wider than the anthropocentrism and androcentrism in which the Jesus Movement was formed.

This is a departure in the history of Christianity and for theology as a whole. We know that the triumphalist, patriarchal attitude in Christianity has projected everything good as having already been present in the Christian tradition, seen as a deep well full of all human richness. This is undoubtedly due to the organization of Christian thought around so-called eternal truths. Today we must go beyond these and the interpretation of Christianity as the repository, even in a latent state, of the truth in all matters. Today we are presented with an invitation from the history of life: the life of different peoples, the life of ecosystems, the life of the planet. The Christian tradition must join a larger movement, beyond the old frontiers, cosmologies, and anthropologies. In re-encountering the wisdom of nature, that of native peoples and marginalized groups, the Christian tradition will learn, as others have, to find a new place for itself in the universe.

This is, then, one of the aspirations of the ecofeminist ethics: to open space to women and the ecosystem to enable them to become rightful subjects in the building of new relationships based on respect and reciprocity. This goal seems also to be part of the wisdom arising from the inability of the economic, political, cultural, and religious patriarchy to promote and respect the rights of human beings and all animals and all sources of life that exist as the body of the Earth.

This goal is a concrete challenge for each and every one of us. We must fill ourselves with it like a person in love, made able by love and mutual attraction to make the world more livable. And finally, we must give praise for being here. Because our being here is mysterious, and that makes a huge difference.

10

Deconstructive Ecofeminism:
A Japanese Critical Interpretation

Masatsugu Maruyama

Introduction

Masatsugu Maruyama, a political theorist, compares what he calls deconstructive ecofeminism with traditional Japanese worldviews, particularly those of Shinto. He shows that the efforts of some ecofeminists to deconstruct dualism and promote an ethics of caring for the natural world, while having some similarities to Shintoism, are not applicable in this religion and context. Although traditional Japanese/Shinto worldviews are not unproblematic from the viewpoint of both ecology and feminism, the difficulties that ecofeminism points to may not be culturally applicable. Masatsugu shows that while the woman/nature connection may apply crossculturally, it is formulated in diverse ways and with varied social consequences. Thus, strategies for change must be context, culture, and religion specific.

From Masatsugu Maruyama, "Deconstructive Ecofeminism: A Japanese Critical Interpretation," *Worldviews* 4, no. 1 (2000). Copyright © 2000 by Brill Academic Publishers, NV. Reprinted by permission.

ECOFEMINISTS ARGUE THAT there is a mutually reinforcing relationship between the devastation of nature and the domination of women in Western society.[1] Two leading ecofeminists, Karen Warren and Val Plumwood, represent one strand of ecofeminism that recommends a kind of philosophical
idealism. They insist that the main cause of this oppression is "hierarchical
dualism," and therefore they seek a strategy of liberation by transcending this
dualism. I call this standpoint "deconstructive ecofeminism," following Dobson's terminology.[2]

Deconstructive ecofeminism of this kind makes three key points. The first
is the *deconstruction of dualism* to which I have just alluded. For example,
Plumwood opposes hierarchical dualism: culture/nature, reason/nature,
male/female, mind/body (nature), reason/emotion, human/nature, and she
develops a counter conception of such dualism.[3] In the same way, Warren
bases ecofeminism on the critique of "value hierarchical thinking"[4] and then
sees the logic of domination, coupled with this value-hierarchical thinking, as
the principal ideological conceptions justifying the domination of women and
nature.[5]

The second point is a *mutual self*. According to Plumwood, the failure of
mainstream environmental philosophies, including those based on deep ecology, is derived from the conception of self in rationalism. This conception
cannot avoid being egoistic because it sees the self against others and does not
leave room for essential connection to others. It instrumentalizes others and
treats nature as an object of the self.[6] Instead of this Plumwood proposes the
conception of a mutual self. In a different context, Warren criticizes liberal
feminism for its extreme individualism. She also recognizes its difficulty particularly in its rights-based ethics. In place of such individualism, she indicates
the interconnectedness of human-nature systems and proposes ecological holistic ethics based on a "web-like" view of relationships among all beings.[7]

The third point is the *care ethic*. This point is logically related to the conception of a mutual self. For example, Warren argues that ecofeminism places
central values on care, love, trust, and mutual reciprocity.[8] In a similar way,
Plumwood maintains that the mutual self stands in a particular relationship
of care, custodianship, and friendship.[9]

These three points are presented by deconstructive ecofeminists as the alternatives leading to nonoppressive society for both nature and women. But
we can find surprisingly similar conceptions within some Japanese traditional
worldviews, particularly those of Japan's indigenous religion, Shinto. In the
doctrines of Shinto, for instance, we find nonhierarchical dualism not only
between men and women but also humans and nature. Shinto also suggests an
awareness of mutual self-concerning between human and human, and moreover, human and nature. And Shinto truly admires care for others, including

natural things. Indeed, although modern developments destroy mountains, hills, and villages, there are strictly defended holy forests around most Shinto shrines; Japanese people treat these areas as sanctuaries. And some Shintoists insist that in the doctrine and practice of Shinto women are given a central position and treated as equal with men.

Yet we also know that Shinto does permit the destruction of nature and assist the oppression of women. For example, Shinto shrines are dedicated even within the "construction companies" that injure wild nature and build an artistic nature in its place. And, in committed Shintoist families, women are segregated even within the family and assigned a limited role irrespective of their endeavors. Importantly, even in these cases, we find the logic that is supposed to end the oppression of women and nature.

If this is the case, can we accept the arguments of the deconstructive ecofeminists? I have been brought to doubt this. To articulate this doubt, I compare their arguments with Japanese indigenous religion, Shinto, speculating how the doctrine of Shinto causes the oppression of women and nature. This will show some difficulties in deconstructive ecofeminism and suggest the need for more refinements in its arguments.

Before suggesting some similarities between deconstructive ecofeminism and a traditional Japanese worldview, some qualifications about Japanese religious philosophies are in order. The first is its unique syncretism. A student of Japanese religion, H. Byron Earhart, once characterized Japanese religion as syncretism. He explains it as if it were a cherry tree. "Japanese religion is a tree, of which the root is Shinto, of which the stems and branches are Confucianism, and on which Buddhism blooms flowers and fruits."[10] This is a somewhat literary expression, but commonly used. In Japan, there exist some strands of religion or philosophy that can be distinguished from others; and there were many real conflicts, particularly between Shinto and Buddhism, during Japanese history. However, they also have influenced each other and now coexist together within diverse cultures. So we cannot completely separate one religion from others.

The second is particularly related to Shinto. What is Shinto? This is a difficult problem. Shinto is an indigenous Japanese religion, but it has no catechism,[11] no founder, and no sacred scriptures. Moreover, representative quasi-sacred scriptures, *Kojiki (Records of Ancient Matters,* 712) and *Nihonsyoki (The Records of Japan,* 720) (known as *Nihongi* in the Western world), were ordered by political authorities for the sake of establishing their state legitimacy. Also, as is now well known, they were strongly influenced by Chinese religions and philosophies. Historically, Japanese ancestors had no written language. So if they wanted to describe their indigenous religion, they had to learn Chinese and to borrow Chinese characters. The complex noun *Shin-to* is a borrowed word, too. The *Shin*, Chinese *Shen*, means *god(dess)* or *divinity*, and the *To*, Chinese *Tao*, means *a way*. Consequently, Shinto was an amalgam of Japanese folk religions, which were also complexes of agricultural cult, nature worship, ancestor worship and shamanism, and Chinese religions and philosophies. But the interpretation of the Shinto worldview has been relatively stable in comparison with other religions.[12] Almost all Japanese have been taught Shinto by their families, communities, or through its special rituals.

In the context of the similarities between deconstructive ecofeminism and Shinto, one crucial feature of deconstructive ecofeminism is undoubtedly its aim to transcend "hierarchical dualism." What is "hierarchical dualism"? Deconstructive ecofeminists do not aim to transcend all types of dualism, or as Plumwood puts it, they do not think every dichotomy results in a dualism.[13] Their target is "hierarchical dualism." Plumwood refers, for example, to culture/nature, reason/nature, male/female, mind/body (nature), reason/emotion (nature), human/nature, public/private, self/other, etc.[14] And Warren, using the work of Elizabeth Dodson Gray, regards men/women, culture/nature, minds/bodies as patriarchal conceptual frameworks.[15] Among these concepts of dualism, the dualism of mind/body or reason/nature is seen as a key.[16] Because they wish to transcend dualism, their alternative is based on a counter conception of this mind/body or reason/nature dualism.

How is this alternative developed? Plumwood discusses several possibilities; the difference between these alternatives is very important for her. What is at stake is not transcendence itself but *how* we transcend. She suggests that one way of overcoming this dualism could be via a mechanistic and reductionistic route. Here, mentality and intentionality retreat from the scene, and instead the conception of the material comes to the fore. In other words, its project is not to resolve the original dualism but simply to reverse it. As Plumwood puts it, this is a "truncated reversal."[17]

Of course, this solution does not satisfy her. First, a mechanistic view sees things as individual beings. Especially when it combines its perspective with the subject/object dualism, it treats objects as separate and independent. But this simultaneously denies dependency between observer and observed and sees the relations between these two actors as quite independent of each other. And second, because mentality was banished, the mechanistic view treats human beings and nature identically. However, by doing so it conceives them to be in a world that is not filled with meaning. This means there is no respect for nature *or* human beings. The lesson Plumwood derives from these observations is that "A deeper resolution of mind/body and associated mind/nature dualisms involves finding a *non-reductionist* basis for recognizing continuity and reclaiming the ground of overlap between nature, the body and the human."[18]

So she then turns to a "non-reductionistic" view as an alternative to dualism. It is taken for granted that such an alternative involves a holistic view like that in deep ecology. But according to Plumwood, the so-called holism on which deep ecology insists does not represent an appropriate way of resolving the problems of human/nature duality.[19] First, holism identifies the problems of the mechanistic view as being atomistic perception, and it is therefore perceived as an alternative to mechanistic dualism. But in her view "atomism does not imply dualism, since ultimate particles may still be treated as having mindlike properties."[20] We do not necessarily have to deny individuality to transcend dualism. And second, the presumption of a simple dichotomy between atomism and holism is in itself false. There are different kinds and degrees of both atomism and holism. Rather, an appropriate "atomism may involve giving a richer account of individuals in nature as both agentic and essentially relational, as well as acknowledging the irreducibility of wholes."[21]

In relation to this argument, Plumwood refers to goddess pantheism, which is sometimes advocated in strands of ecofeminism. Here, too, she finds difficulties. In pantheism nature is perceived as being perfectly sentient and thought of "as human." Thus, there is no recognition of the difference between human and nonhuman. Nature is anthropomorphized, grasped through the model of human nature—an anthropocentric point of view. Although divinity can be thought of as immanent in all things, this divinity is often seen as sharing with every thing and as connected with the Whole or One. If this is the case, the individual will be seen as a puppet of a central principle. In short, "A form of pantheism which recognizes neither difference nor diversity, and in which each particular is respected only for its sameness or goddess component, remains within a hierarchical and dualistic framework."[22]

However, the simple recognition of individuality does not constitute an alternative. It inevitably leads to a subject/object dualism. To avoid this, something common to all—human and nonhuman—needs to be identified. For this, Plumwood focuses on varieties of mind and chooses intentionality.[23] In her judgment, scientific projects designed to establish a clear separation between human and nonhuman do not succeed in denying wide-ranging intentionality. And, "Once mindfulness is conceived in more diverse, continuous and graduated ways, the failure of discontinuity leads only to the much more plausible thesis of weak panpsychism, the view that *mindlike* qualities are to be found in nature."[24]

Now we can sum up Plumwood's arguments. She identifies an ecofeminist liberation strategy as the changing of our cognitive map of the world. She rejects mechanism, hyper-atomism, extreme holism, and pantheism as containing a hidden oneness. She is, however, prepared to endorse weak holism, weak atomism, and weak panpsychism. And in her opinion, such an alternative worldview can produce the perception of continuity and difference between humanity and nature.

Let us turn to the Shinto worldview, where I begin by reviewing some parts of its quasi-sacred scripture, *Kojiki*:

> The names of the Deities (Kami) that were born (or became) in the Plain of High Heaven when the Heaven and the Earth began were Ame-no-minaka-nushi-no-kami (The Deity Master of the August Center of Heaven), next Taka-mi-musubi-no-kami (The High August Producing Wondrous Deity), next Kami-musubi-no-kami (The Divine Producing Wondrous Deity). These three Kami were all Kami born alone, and hid their persons.[25]

This is a starting point. In Shinto, Kamis are plural and their sexes are indefinite at least in this part because they were born alone or matured alone. But sexuality appears shortly after:

> Next Izanagi-no-kami (Deity the Male who Invites), next his younger sister Izanami-no-kami (Deity the Female who invites [sic.]). . . .[26] All the other Heavenly Kami . . . collectively commanded them to "make, consolidate and give birth to this drifting land." . . .[27] Having descended from Heaven onto this land, they saw to the erection of a heaven august pillar. . . . Then Izanagi asked his younger sister Izanami: "How is your body made?" She replied: "My body grew growing, but there is one part of it which has not grown continuous." Izanagi then said, "My body grew growing, but there is one part of which grew superfluous. Would it not be good that I should insert the part of this my body which grew superfluous in the part of your body which did not grow continuous, and that we should procreate regions?" In reply, Izanami said: "It would be good."[28]

In this quotation, a god and goddess appear and they are married.[29] The Kamis produce, first, many islands, lands, and lakes, which of course make up the land of Japan, and then many Kamis, which include the Kami of ocean, Kami of wind, Kami of trees, Kami of mountains, Kami of birds, and so on. Moreover, when Izanagi washes his body after visiting "Yo-mi (darkness)" country to meet "dead" Izanami,[30] many important Kamis are born, in which there is an Ama-terasu-oo-mi-kami (Heaven-shining Great August Goddess, or the Sun Goddess).[31] She is commissioned to rule the Plain of High Heaven by Izanagi and decides to send her grandson, Ninigi, to rule Japan. Of course, in this book, this ruler is described as the ancestor of the present emperor, Ten-no (Heavenly Emperor). On this reading, the Japanese emperor is a descendent of Kami, particularly of the Sun Goddess, and he or she is also perceived as Kami itself, Ara-hito-gami (this-worldly Human God).[32] In addition to this, in *Kojiki*, humans appear without notice and sometimes are married Kami (of course, other Kamis are born from these marriages).

From these materials, we can construct some features of the worldview in *Kojiki*. First, as in many eastern philosophies, in Shinto, there is no mind/body hierarchical dualism. There is, however, a belief that something eternal can survive beyond natural death. It is called *Tama, Mono,* or *Mi.* (All these are quite similar to the English *soul,* but they are distinguished from each other. Tama can be revered as Kami, but other souls do not have such possibilities.) People believed that they sometimes left the body even during its lifetime. But people did not see Tama as governing the body; rather they saw it as living within the body also. Not only animals but also plants and minerals were seen to have their own Tama. And interestingly, people distinguished some kinds or aspects of Tama, although the wholeness of Tama was held to be constant. The number and their functions are controversial, but there are at least two discernible facets of Tama. One is called *Ara-mi-tama* (in this case the word *mi* is an honorific prefix attaching to the respected noun), and another is *Nigi-mi-tama.* According to most scholars of Shinto, the Ara-mi-tama is "variously defined as wild, raging, raw, the power destructive of what is evil and constructive of what is Divine, a spirit empowered to rule with authority, manifestation as opposed to essence."[33] On the other hand, the Nigi-mi-tama is believed to be a counterpart of the Ara-mi-tama, and is "described as mild, quiet, refined, peace, what gives peace, what makes adjustments to maintain harmony, a spirit empowered to lead to union and harmony, essence as opposed to manifestation. It is sometimes stressed that it is the essential and original part or aspect of the whole *mitama.*"[34] I think this contrast shows the influence of the yin/yang conception on Shinto. But for the moment, the point is that Shinto does not fundamentally distinguish between Kami, human, and nonhuman. As described in *Kojiki,* in Shinto, in terms of their origins, humans and nature are blood relatives of Kami. They have different appearances but they have a common essence.

Second, the worship of the Sun Goddess might be predominant in Japan because, according to *Kojiki,* she sent the ruler to Japan. But this is not the case. Amaterasu was never seen as the head of pantheons of Kami. However, from the Meiji period to World War II (1868–1945), which is known as the restoration of the emperor, because of the state ideology, this mythology was emphasized by the authorities. To legitimize the power of the emperor, the mythical line of descent from Amaterasu had to be stressed. However, generally, throughout Japanese history, the worship of Amaterasu did not occupy a predominant position in the Shinto religion because people believed that each Kami had a function proper to its character. For example, in Shinto, each family worships its ancestral Kami, household Kami, and Kami of community. People ordinarily worship these Kamis as their closest Kami. And, for example, if they want to improve their ability to learn, they will go to the

Ten-man-gu shrines that were constructed for honoring the famous scholar, Michizane Sugawara (845–903), as the Kami of Literature. All in all, no one Kami can pretend to be a sovereign god(dess).

Third, as we see below, the above features apparently show that Shinto is polytheistic, and moreover, there is neither an absolute hidden Oneness nor a hierarchical structure. But at the same time, this does not necessarily exclude the presence of a principle that we can regard as an integrating principle. That is to say, in the Shinto religion there exists not only weak atomism but also weak holism. Why? The point relates to the name of Shinto itself. As I said above, the complex noun, *Shin-to*, was made from two ideograms. In Japan, many Chinese ideograms have two pronunciations, that is, a Chinese pronunciation (but slightly different from the original) and a Japanese pronunciation. *Shinto* is a Chinese pronunciation, and in such a case, it means *the way of god(dess)* or *divinity*. But if we read it in Japanese, it is pronounced *Kannagara (Kami-nagara)-no-michi*. And, importantly, in this case, it means *to follow the way of Kami*. This perception spiritually guides the Japanese and it also constitutes the ethics of Shinto. I outline its contents and its character below. For the moment, what I want to stress is that this perception of the way of Kami prescribed the definite behavior of humans and it could give cohesiveness to Shinto religion.

The second common feature of ecofeminist deconstructivist discourses is a critique of the egoistic self in liberal individualism. They suggest that the liberal self is clearly demarcated from other selves and cannot help treating others as instruments for itself to attain its egoistic goals. This instrumental, separated conception of the self, they argue, does not give an appropriate image of the human self because humans are social and connected beings. Simultaneously it eliminates basic insight into gender difference.[35]

In the place of this conception of the egoistic self, deconstructivist ecofeminists propose the relational type of self. Plumwood develops this further. According to her, the relational self can offer a noninstrumental mode of perception that grounds a new ethics (explained in detail below), and which asks us to treat others for themselves.[36] She aims not only to recognize kinship but also difference. In order to carry out this double strategy, she advocates the idea of the "mutual self." Here, the key point is intersubjective interaction. She explains this category through referring to the book, *The Bonds of Love*, written by Jessica Benjamin:

> For Benjamin, the process of mutual transformation or recognition, the "dance of interaction," is the basis of the formation of self through mutuality, a process in which an external other sets a boundary or limit to the self and its desires. This formation involves the recognition of the other as alike (non-alien) but as different, as other.[37]

The mutual self admits relationship with others and simultaneously recognizes the "earth other" as an agent having intentionality and its place, and moreover understands the other as setting limits on "right" action.

Now let us turn again to the Japanese case. Plumwood speaks of two types of mutual self, "the truly social self" and the "ecological self;"[38] therefore, I point out the first in the self-perception of Japanese worldviews and the second in Shinto.

As for the "true social self as mutual self," I think this conception of the self is similar to that known as the "Japanese self." There are many scholars who recognize this self. Linguistics shows that the Japanese language often does not use the subject in a sentence. Nor does it distinguish the subject from the object. Japanese people often do not mind what the subject is in one sentence and often infer the indicated subject from the referred environment.[39] These features weaken the sense of subjects and determine words according to not subject/object itself but the situation that surrounds the subject/object. Compared to this, in English there is a very definite and strict separation between subject and object. The Japanese social anthropologist Nakane contrasts the difference between the Western view of self/other relationship and the Japanese view. According to her, in the Western view, the self and other are seen as poles, while, in the Japanese view, the self and other are interconnected with one another.[40] Fundamentally, the Japanese word *nin-gen* (*human*) consists of two terms. *Nin* means *person* or *people*, and *gen* means *space, space between,* or *relationship.* In Japanese *human* means *the person who stands between other humans.*

Then how about the "ecological self"? As pointed out before, in Shinto, *human* and *nature* are seen as blood relatives. Therefore, an opposition between human and nonhuman does not exist. Both human-derived Kami and nature-derived Kami are revered indiscriminately. But how far does this perception influence the Japanese view of nature? We cannot say exactly the degree of this influence, but there is evidence for such an influence. I will suggest only two cases. The first is old poems that were sung in the same time as the *Kojiki.* These Japanese old poems show how people often preferred to express their intentions through identifying themselves with nonhuman beings. For example, Hitomaro Kakinomoto, one of the representative poets in *Manyosyu,* writes "O you yellow leaves, That fall upon the autumn slopes—If only for a moment, Do not whirl down in such confusion, That I may see where my beloved dwells."[41] This type of expression is called *Mono-ni-yosete-omoi-wo-nobu.* This can generally be translated as *an expression via nature.* In this case, *Mono* means *nature,* but interestingly, as I have suggested before, *Mono* means simultaneously *soul.* In ancient times, Japanese people saw nature not as "dead things" but as "living things" even if it did not scientifically have a life. So at this time it was thought that all natural beings as well as humans have selves in their own manner.

The second case of influence from Shinto is the transformation of Buddhism that is well known as *Japanizing*. We can trace many cases of this, but I mention the discussion on Dogen by Curtin because this case directly shows the perception of ecological self.[42] Curtin stresses the difference of self-perception between that of deep ecology and that of the Japanese Zen founder Dogen (1200–1253), although Dogen has influenced many deep ecologists. According to Curtin, for Dogen, "the self does not disappear or merge into the cosmos. He never denies that there are multiple, provisional, contextually defined borders that shape the sense of self. . . . Self is always experienced *in relation* to other beings, however, and those relations define what it means to be a self."[43] Curtin cites this passage as evidence for his interpretation: "When you ride in a boat, your body and mind and the environs together are the undivided activity of the boat. The entire earth and the entire sky are both the undivided activity of the boat."[44] And "Although not one, not different; although not different, not the same; although not the same, not many."[45]

As Curtin rightly points out, there is here certainly a sense of the particularity of the self as well as the relationality of self to earth other. Curtin also highlights Dogen's unique interpretation of *Bu-ssho* (*Buddha-nature*). In traditional Buddhism, living things are seen to have the potentiality of enlightenment because all of them immanently have this Buddha-nature. But there is so deep a gulf between a normal living thing and an enlightened thing that such a reading allows for dualism. Dogen denies this dualism and reaches the standpoint of absolute equality among all living things. Curtin explains this point particularly by focusing on Dogen's intentional misreading of the sutras. Curtin remarks that "Dogen twists the expression 'All sentient beings without exception *have* the Buddha-nature' to read 'All sentient beings without exception *are* Buddha-nature.'"[46] According to Curtin, on the former reading, Buddha-nature remains a possibility and this allows discrimination against the nonenlightened, but in the latter reading, there is no exception for the enlightened and this ensures the rule of equality among sentient beings. And Curtin insists that Dogen also enlarges the scope of Buddha-nature; that is, he admits Buddha-nature in nonsentient beings. Curtin argues that it is this that makes Dogen's ecological thought unique.

I am not sure whether Curtin's interpretation of Dogen's Buddhism is accurate or not. Curtin insists on the uniqueness of Dogen's view of nature, particularly Buddha-nature. However, in Japan, debates about Buddha-nature in relation to ecology develop rather around the *Hon-gaku* (*nature of enlightenment*) thought. This thought is said to come from the Chinese Buddhist Tannen (711–782). He taught that nonsentient beings also (not exclusively sentient beings) have Buddha-nature. His teachings were transported into Japan in the ninth century and these gained authority in many sects of Japanese Buddhism. And importantly, *Hon-gaku* thought is interpreted as suggesting that all beings (literally speaking in Japanese, "mountains, rivers, grasses and plants") should attain enlightenment.[47] This is apparently deviant from original Buddhism.

Of course, in both the cases of Dogen and of *Hon-gaku* thought, we cannot attribute such deviation to Shinto alone. It may be immanent in original Buddhism. Rather, the point is another; if we turn our attention to Shinto, we need not be surprised at Dogen's arguments about Buddha-nature, let alone Hon-gaku thought. From the standpoint of Shinto, the interpretation of individual Buddha-nature as immanent reality in all individual natural beings is literally natural because we can accept individual Tama as Buddha-nature.

In general, those who advocate the relational self from an ecofeminist standpoint also insist on an ethic of care as a connected concept of the self. For example, Jim Cheney remarks that care for particular others develops through given rather than abstract connections where individuals are seen as a part of a web of relations.[48] This is also true of deconstructivist ecofeminists. Warren mentions the difference between climbing with an arrogant eye and with a loving eye and illustrates the latter by the climber who takes care for a rock as embodying the ethic of care.[49] In doing so, and referring to Cheney, she concludes, "Ecofeminism makes a central place for values of care, love, trust, and appropriate reciprocity—values that presuppose that our relationships to others are central to our understanding of who we are."[50] In a similar way, Plumwood argues: "The ecological self can be interpreted as a form of mutual selfhood in which the self makes essential connection to earth others, and hence as a product of a certain sort of relational identity. . . . He or she stands in particular relations, which may be those of care, custodianship, friendship, or various diverse virtue concepts."[51] In the relationship of the mutual self, the other is treated as an object to be cared for. Conversely, if we do not treat the "other" as something to be cared for, our relationship to the other cannot be regarded as deriving from a notion of the mutual self. "Care" and "mutual self" cannot be separated.

Here we should be careful of the qualification about the conception of caring because, even under "hyper-individualism," an ethic of care is sometimes advocated. To discuss this point, we should look at the original conception of care ethics used by deconstructivists[52]—that of Carol Gilligan.

Based on empirical observations, she identified two different orientations in moral judgment: justice and care. According to Gilligan, these are not opposed, but rather each takes a different departure point for moral consideration and also the former is more associated with the male and the latter with the female. While the former starts from a distinction between self and other, the latter starts from a relationship between them. And importantly, "With *the shift in perspective from justice to care* . . . words connoting relationship like 'dependence' or 'responsibility' or even moral terms such as 'fairness' and 'care' take on *different meanings*."⁵³ As this citation shows, she insists, if we take a care perspective, we see moral problems quite differently from a justice perspective. From the care perspective, it is not the self but the relationship that comes to the fore. The self does not judge the conflict situation between it and others from a rule of equality; rather it is the relationship that defines what should be done. The moral question shifts from "What is just?" as manifested in the Categorical Imperative or the Golden Rule to "How to respond?"⁵⁴ According to Gilligan, these perspectives are theoretically not opposite or mirror-images of one another, but empirically "people have a tendency to lose sight of one moral perspective in arriving at moral decision—a liability equally shared by both sexes."⁵⁵ When deconstructivists refer to "care," we should keep in mind the connotation of care described above.

In the Japanese religious context, we can find many elements of an ethic of care. Indeed we can even say such doctrines are abundantly present in all of the Japanese religions; but here, I refer only to the teaching about *Makoto* in Shinto.

It has been pointed out that Shinto has no recognized ethical code like the Ten Commandments. Many Western Christian missionaries have historically looked down on Shinto for this reason. But Herbert says, borrowing from a Japanese Shintoist, that Shinto "tries 'to deal with the sphere of man's problem which is beyond the sphere of good or bad conduct. . . . The problem of how to emancipate man from worries and anxieties [and] in Japan the task of religion has been concentrated on the latter problem,' i.e. 'the inner problem of man,' not 'preparing ethical principles for man's social conduct.'"⁵⁶ In Gilligan's terms, in Shinto, the ethical problem is not "What is justice?" but simply "How to respond to our situation?" "*Makoto*" is a reply to this question.

In general, *Makoto* means *sincerity* in English. But to perceive it more appropriately in Shinto, we need to understand the meaning of the purification of mind and body. Shinto is profoundly concerned with purification.⁵⁷ Almost all rituals in Shinto are related to purification. Rituals are needed to cleanse the pollution caused by death or childbearing, and some rituals are oriented toward attaining the purification of mind and body. In line with this purification, *Makoto* is seen as one of the verifications of mind purity, and in this case *Makoto* is dedicated to the Kami of others. That is to say, the preservation of *Makoto* is a religious endeavor, and so we might even say that "It is by serving the Kami with *makoto* that man can conform to the will of the Kami."⁵⁸

If we understand this point, we can more easily understand why Japanese people conform to the teaching of *Go-rin* (Five ethics of human relationship) in Confucianism. In this teaching, people are taught that one has without exception relationships with others such as husband or wife, children or parents, brother or sister, friend and friend, and lord and retainer. In a traditional society, these five relationships cover all human conditions in a society. In other words, if people want to practice their *Makoto*, they do it in line with the teaching of *Go-rin*. Moreover, although Confucius explained in various ways the meaning of *Jin* (*benevolence*) that is cardinal in Confucianism, one of them was in the form of *Chyu-jo*. And *Chyu-jo* means sincerity and sympathy. As for the responses to other humans, Shinto and Confucianism coincide with each other. Both pay attention to emotions and their ethics lean more toward concrete love in relationships than to general justice. I think these points are quite similar to deconstructivism's new ethics of care.

In line with the argument for the relational self, Plumwood describes how instructive non-Western aboriginal culture is in guiding ecofeminist principle.[59] Her alternatives are undoubtedly based on aboriginal culture, and she thinks this can provide the foundations of a new ecofeminist worldview, although she admits a difficulty with converting from Western culture to a non-Western one in the West. However, is such a worldview unproblematic? I consider this question in the context of criticisms of Japanese culture, especially of Shinto. I begin with human-nature relationships, then consider man-woman relationships.

As outlined above, Plumwood relies on the concept of intentionality to ground the continuity and difference between humans and nature. She argues that even plants have intentionality, although the degree of intention is different from, for example, sentient beings.[60] Moreover, she insists that we should admit a kind of teleology in order to transcend mind/nature dualism and to obtain alternatives to a mechanistic worldview. Of course, this teleology should neither be anthropocentric Aristotelian teleology nor an anthropomorphically animistic one. Rather, we should admit there are many kinds of teleological concepts.[61] I suspect some nihilism here, which I explain by connecting it with historical relativism.

Maruyama points out that in the mythology of *Kojiki* and *Nihongi*, we can find a standardized view of history,[62] summarized as *tugitugini-nariyuku-ikioi* (continuously bearing, transforming, or completing energy).[63] He explains this by comparing it to the ideology of reactionism and progressivism. On the one hand, under reactionism, the ideal society is given by ancient history, and current history is judged by this ideal standard. On the other hand, under progressivism, the ideal society is set by the aim of future history and the present is seen as one step in this ideal direction.[64] His discussions are informative in making clear Shinto's perception of history. The worldview of Shinto is, certainly, in contrast with both.

On the former reactionist reading, current time is judged by old honorable events. If the present course deviates from the old one, it will be accused in the name of justice. But as we saw before, in Shinto, justice is in the background. Judging what happens according to human standards, presupposes that humans are capable of making such judgments about existence. This human arrogance is denied in Shinto. It sticks to a "value-free" position, or more precisely, it refuses judgment itself. Because Kamis always live present lives during history[65] and express their way through every event, we need only to accept ongoing reality and should not dare to see it from past golden times.

On the latter progressivist reading, an ultimate goal or end point along the historical line in the future is presupposed. Every human history is seen as running through a similar course even if it currently appears to be in a different stage. And the final stage is the product of human speculation. But in Shinto, everything blossoms according to its own reason. Certainly, this presupposes a kind of logic or reason, but this reason is pluralistic. It does not dare to examine the logic. It only accepts reality as it is given. If we try to predict, this would show our arrogance and would amount to the sin of disrespect for Kami. Ueda explains the crucial difference here: "the essence and value of existence is to be discovered not in an absolute, *a priori* rational principle (Greek, *logos*), nor again in a universal norm or law (Sanskrit, *dharma*), but in the possibilities inherent in concrete forms of existence. Accordingly, Shinto is a religion of the relative—in the positive sense that it is committed to reality in the endless process of becoming. It is for this reason that Shinto places such a high value on the birth of new life and on the transmission of life through successive generations."[66]

Historical relativism thus permeates every discourse in Shinto. We can infer from this that the perception advocated by Plumwood—that intentionality is present in all beings—may also lead to historical relativism. We cannot help accepting all events even if they are problematic from the viewpoint of our human reason.

We can also identify another problem in the continuity of nature and culture. King points out that ecofeminists do not clarify what caring about nature involves; that is, whether it is perceived from an essentialist standpoint (i.e., women are biologically closer to nature than men, so they necessarily care more about nature) or from a conceptualist standpoint (i.e., a dualistic conception of men/women, culture/nature underlies a patriarchal culture, so a new ethic that has been ignored by the predominant culture is needed).[67] Clarifying the latter point, he takes up the short story of rock climbing with loving eyes by Warren herself, which she considers as illustrating ecofeminism's ethic. In the context of this story, King poses a question about its strong leaning toward lived experience: "lived experience is selective in that it results from cultural, as well as personal, interpretation of experience."[68] Since there are so many people, there are many lived experiences. Thus, many voices about the relationship between human and nature are heard. The lived experiences of farmers, hunters, tourists, and city dwellers are very different from each other. How can we decide which voice expresses the care ethic? Thus King doubts the extraction of care ethic from lived experience. But I think the problem can be viewed in another way. If various people insist on different care relationships to nature according to their interpretations, how do we judge which is the appropriate care ethic?

Here, Callicott is quite suggestive (although he himself does not address this problem, being more concerned with deep ecological elements in Japanese religions).[69] He collects the discussions of scholars who have argued that the Japanese traditional conception of nature was heavily vested with cultural ingredients and that therefore it was totally different from Western conceptions of nature, particularly that of wilderness. For example, Grapard points to the dialectical relationship between nature and culture in Shinto and Japanese Buddhism, and in his view, "it might be said that what has been termed the 'Japanese love of nature' is actually the 'Japanese love of cultural transformations and purifications[70] of a world which, if left alone, simply decays.' So that *the love of culture takes in Japan the form of a love of nature*."[71] Likewise, Totman rejects the explanation that, due to the love of nature in Japan, many forests have survived even in densely populated places, and insists that the nature celebrated in traditional Japanese culture "is an aesthetic abstraction that has little to do with the 'nature' of a real ecosystem. The sensibility associated with raising *bonsai*, viewing cherry blossoms, nurturing disciplined ornamental gardens, treasuring painted landscapes and admiring chrysanthemums is an entirely different order of things from the concerns and feelings involved in policing woodlands and planting trees."[72] French Japanologist Augustin Berque reinforced these opinions. He stated that, "As is well known, Japanese culture has paid delicate attention to its natural environment; but this was *not environment in general*; it was a selection of some places . . . some plants . . . some moments of the year . . . etc., *all entangled into certain sets of regular associations*."[73]

I think these arguments point to some common features of the Japanese love of nature. First, as Callicott says, it amalgamates nature and culture. There is no perception of a contradiction or separation between culture and nature nor that the nature that people care for is a result of humans intervening in and transforming nature. This may give an unlimited excuse for intervening for the sake of nature itself. Second, the love of nature is selective,[74] sometimes leading to a culturally defined selection of "natural" elements. Third, and closely related to the second feature, is its disinterestedness in the totality of nature. Caring does not mean nontouching or leaving alone. Rather, caring requires much attention and constant labor. But caring for particular natural objects may cause an imbalance in the wider ecosystem.

What is the origin of these difficulties? I believe the key lies in the care ethic. To argue this point, I shall lastly discuss problems with the care ethic itself.

King points out a problem of the care ethic in Warren's account. Warren, he says, describes care as a subjective feeling linked to an awareness of experience. She feels her friendship toward the rock. She even says that she experiences a feeling of conversational partnership even though the rock cannot talk. But how can we hear the voice of rock? When Warren says that she communicates with rock, she seems only to hear what she wants.[75]

This problem of subjectivity is similar to that referred to in the Japanese perception of human-nature relationships in the previous section. But King adds another problem: "She [Warren] contrasts a climber who cares about the rock with one who seeks to conquer it, yet for the rock, it is all the same thing; the rock does not care. Indeed, the fact that the climber cares for the rock appears to have no *practical consequences* for the rock itself."[76] King concludes that there is an ambiguity in the conception of care within "conceptualist" ecofeminism. This criticism is very important, but it is not the case that the meaning of care about nature in deconstructivism is unclear. On the contrary, it has a very clear character.

In arguing this point, the two ideal action types of ethic presented by Max Weber are informative. According to Weber, "ethically oriented activity can follow two fundamentally different, irreconcilably opposed maxims. It can follow the 'ethic of principled conviction (*Gesinnung*)' or the 'ethic of responsibility.'"[77] Of course, the former does not mean irresponsibility, nor does the latter exclude conviction. They are not opposing poles. But there is a great difference about which kinds of things the action cares for. In the former, if we put it in a religious form, "The Christian does what is right and places the outcome in God's hands."[78] That is to say, in the course of action relating to a conviction ethic, the actor pays more attention to his or her sincerity. He or she does not doubt this firm belief and always has to feel this conviction. In the latter case, the actor pays more attention to the *consequences* caused by his or her action. Of course, we cannot foresee everything, and we often produce results that we do not intend. But the point is not the possibility of foresight, but our regard for *consequences*.

In the light of these two ideal types, which type is closer to the care ethic in deconstructivism?

In a paper presented in 1997, "Situated Universalism and Care-sensitive Ethics," Warren admits that even among "care theorists" there are great disagreements about "whether an 'ethic of care' is compatible with, distinct and independent from, or more basic than, 'an ethic of justice.'"[79] She does not show her own hand in these debates, rather she wants to move a point and focus on the standpoint that all "care theorists" seem to believe in as the minimum condition for ethical deliberation. According to her, this condition can be summarized so: "One *cannot* do ethics . . . *unless one cares.*"[80] That is to say, she insists that every ethical position needs care as a minimal moral requirement, and therefore all ethical attitudes presuppose "care-sensitive ethics."

This argument mainly depends on the scientific research by Daniel Goleman. According to Goleman, "there are two 'minds' or 'brains,' 'one that thinks and one that feels,' . . . rational minds ('reason') and emotional minds ('emotion'). . . . They provide two different kinds of intelligences." And moreover, "Goleman claims that 'emotional intelligence' . . . is what maintains an appropriate balance between the two 'minds.'"[81] Warren draws her thesis from this "scientific knowledge."[82]

Of course, as she rejects hierarchical thinking, she does not insist that the "emotional mind" has or should have the dominant position in total intelligence. Rather, she states that emotional and rational intelligence are "in concert" when we act ethically.[83] But if a conductor stresses some parts compared to others, or if some musicians miss a beat, or a particularly emotional part comes to the fore, what consequences will it have? It was these types of problems within a *political* context with which Weber was concerned. (His essay on this subject is called "The Profession and Vocation of Politics.") So long as we see environmental hazards as a *social* and hence *political* problem, we cannot underestimate his warning.

Now I turn to the relationship between man and woman. In general, ecofeminist deconstructivists say plenty about human-nature relations but, curiously, they scarcely articulate how they, as feminists, think about the relationship between men and women. What we can guess is that they are eager to arrive at something different from the relation advocated by liberal feminism, radical (cultural) feminism, Marxist feminism, and socialist feminism.[84] Perhaps the concept of a "relationship of non-hierarchical difference" presented by Plumwood is most common.[85] She explains this by presenting a counterargument against dualism. According to her, it will have the following features: the denial of, backgrounding, radical exclusion, incorporation, instrumentalism, and homogenization.[86]

I do not think, as Plumwood does, that these features are so different from the concept of "individual dignity" in liberalism. But since she sees that liberal feminism aims for "uncritical equality" and tries to fit "women to a masculine model," the main point for her is not "merging" into the liberalistic ideal but achieving "non-hierarchical difference." If this principle is then applied to male-female relationships, how does it appear? At this point I want to turn to look at Shinto.

While Confucianism and Buddhism in Japan are often thought of as sexist, it is much more difficult to locate sexism in the discourses of Shinto. On the contrary, since it is predominantly a religion of rice farming, it reveres women as having supernatural powers of birth, and it admits the superiority of *Nigi-mi-tama* that has female features against *Ara-mi-tama* that has a male character. Japanese emperors, who are said to be descendants of the Sun Goddess, are often regarded by people as mother, not father, even if they are men. So, in the famous 1911 manuscript of the feminist movement in Japan, when the author, Raichou Hiratuka, declared that "In ancient times, woman was really the sun," the phrase did not seem so absurd. In fact, many priestesses, including high ones, are still active even now in Shinto. But one exception exists. As I mentioned before, Shinto pays great attention to *Kegare* (*pollution*), and in this category menstruation and childbearing are included, and these are known as two of the three fundamental categories of pollution.[87] Women are sometimes and in some places excluded for these reasons. However, Shinto allows us to see pollution nonessentialistically.[88] A ritual can wipe out pollution or it can change into cleanliness after a prescribed period. It is fundamentally different from the perception of original sin immanent in humanity. So at least Shinto admits the contribution of women, affirms the continuity of man and woman (everything has an element of yin and yang), and recognizes that everybody should enjoy his or her life.

In the light of this, one might anticipate a feminist society in a Shinto world. Unfortunately, we do not know how society was constituted under "pure Shinto" since written history in Japan (as Western history) almost always does not refer to women. This is in itself a problem: is this masculine history written in spite of the teachings of religions or as precisely because of them? To illuminate this point, we should focus on the care ethic under patriarchy.

Cuomo has criticized the care ethic advocated by Warren.[89] Cuomo argues that caring does not necessarily represent the morally good and can be morally harmful where it ignores other responsibilities. More crucially, she argues, "In fact, *female caring and compassion for oppressors* are cornerstones of patriarchal systems. Women have forgiven oppressors, stayed with abusive husbands and partners, and sacrificed their own desires *because of their great ability to care for others.*"[90]

So, if a religion stresses the teaching of the care ethic, it would recommend that a person showed more caring rather than less, admire the person caring for others irrespective of whether the others deserved care or not, and sometimes praise that person like a Kami or near-Buddha. The consequences of doing so are to assist the reproduction of the social system. It is this role that can be performed by Japanese religions.

In Shinto, for instance, the teaching of the ethic about *Makoto* supports this attitude. It taught that we should show *Makoto* to objects irrespective of their goodness or badness. Moreover, "The attitude of *makoto* is considered to be that which leads to *wa*."[91] Although we cannot exactly define this word, *Wa*, in English, it means *harmony* as well as *peace*. In the Seventeen-article Constitution supposedly made by Shotoku, he gave the first article as, "You should venerate Wa." In general, it was said that this thought came from the *Analects* of Confucius, but a Japanese scholar of Buddhism insisted that it could be attributed to *Ji-hi* (*compassion*) in Buddhism if it was compared with the other articles.[92] Its origin is not important. What is significant is that Japanese religions can insist on this ethic, which is at the same time seen as a principle of natural beings as well as human beings. And this idea of harmony and peace—both of which would be advocated by ecofeminists—directly reinforced the status quo in Japan. Shotoku also noted in the Seventeen-article Constitution that "You should not try to resist." If this rule were also revered, a woman trying to resist the patriarchal social order would be not only accused by law but also blamed by the religion.

Finally, I discuss Shinto's "relationalism." As we saw above, in Shinto everything including nature can be seen as related, with a common origin. Humans are perceived first as relational beings—a member of a family and a community, which includes the land, clan, and nation. Therefore people revere the Kami of family, community, clan, and nation. No human can survive forever. But as he or she has a vertical interrelationship with ancestors and descendants and a horizontal interrelationship with many degrees of community and natural beings, life itself continues through birth and death. According to such a perception, how do people behave? A Japanese scholar of Shinto describes it so:

> In Shinto it is believed that diligent endeavor in daily life is at once the will of the *kami* who brought this land into being and the fulfillment of their will. The *kami* bestowed their blessings on the people of this land and desire that men's lives should never cease to be productive and fruitful. To accept these blessings and live in the awareness of one's identity as a member of an entire people is to live as a *mikoto mochi*,[93] a person who mediates the will of the *kami* through his daily life. Individual existence finds fulfillment only as one realizes that one's own life is inseparably bound up with the task of passing on and enhancing the life of the people one belongs to.[94]

In this passage, the author expresses *human* as *male*, but this is the result of English translation; Japanese does not have a sexually biased word for human. The way of life in Shinto, *mikoto mochi*, is applied to both sexes. Then what is the role of woman? The answer is simple. She is viewed as a mother. Since

Shinto conceives a reverence for the reproduction of life, the task relating to birth, nursing, and caring is viewed as a sacred task. What woman has to do is to comply with this mothering role. Japanese women were taught that this sexual role was a holy duty. This is the position closest to essentialist cultural ecofeminism.

As we saw above, deconstructivist feminism departs from this standpoint because of its dualistic worldview. But when we see that Plumwood explains the relational self in terms of a mother's care for her child, and states "She [Mother] wants health and happiness *for* the child for its sake, as well as for her own,"[95] or that she admits "an important ground of certain caring relations would be a locally particularized identity involving commitment to a particular place and its non-human as well as its human inhabitants,"[96] I suspect that she plays the same role of a priestess who advocated the way of woman in Shinto. In other words, practically, Plumwood in this respect is very close to those Shinto priestesses who advocate the way of women as mothers and child rearers. This is quite the opposite to what she intends to achieve through deconstructive ecofeminism.

In this essay I discuss two theses. The first one is that there are surprising similarities between the discourses of deconstructive ecofeminism and traditional Japanese religions, especially Shinto. The second is that the alternative presented by these ecofeminists might have serious difficulties if the first thesis is correct. For traditional Japanese religions are not unproblematic. The culture of nature-love does not ensure the love of nature. Love may not only unintentionally hurt nature but may also intentionally remake nature regardless of ecological rationality. And the male-female relationship that follows the principle of deconstructive dualism might not necessarily lead to equality of men and women.

But more importantly, it is the teaching of these religions that almost all current Japanese feminists criticize. Moreover, they have to fight against "mother discourses" because the ideal of "mother" historically consisted of the prewar authoritative state ideology and, even now, it backs up the social system that has been maintained by "corporate warriors." The Japanese case also suggests that hierarchical dualism is not the only underpinning of the devastation of nature and women's oppression, since it is not present in Japanese religion; neither are nature and women identified with one another.

Against such criticism, deconstructive ecofeminists could defend their position by saying "at least in the Western world" in relation to their fundamental perception of nature-women oppression. And they admit the cultural plurality of feminism. However, if an oppressor were to welcome deconstructive ecofeminist messages, are they assisting the oppressor unintentionally? Deconstructivist ecofeminists have the freedom to seek for a utopia in an exotic culture, but they do not have a right to unintentionally maintain the patriarchal sexist structure and ideology within other cultures. Since they pay more attention to intentionality, they are inclined to lose sight of problems associated with their *own* unintentionality.

Notes

1. Greta Gaard and Lori Gruen, "Ecofeminism: Toward Global Justice and Planetary Health," *Society and Nature* 2, 1 (1993): 5.

2. Andrew Dobson, *Green Political Thought*, 2d ed. (London: Routledge, 1995), 169.

3. Val Plumwood, *Feminism and the Mastery of Nature* (London: Routledge, 1993).

4. Karen J. Warren, "Feminism and Ecology: Making Connections," *Environmental Ethics* 9, 1 (1987): 6.

5. Karen J. Warren, "The Power and the Promise of Ecological Feminism," in *Ecological Feminist Philosophies*, ed. Karen J. Warren (Bloomington: Indiana University Press, 1996), 20.

6. Val Plumwood, "Nature, Self, and Gender: Feminism, Environmental Philosophy, and the Critique of Rationalism," in *Ecological Feminist Philosophies*, ed. Karen J. Warren (Bloomington: Indiana University Press, 1996), 170.

7. Warren, "Feminism and Ecology: Making Connections," 10.

8. Warren, "The Power and the Promise of Ecological Feminism," 33.

9. Plumwood, *Feminism and the Mastery of Nature*, 184–85.

10. H. Byron Earhart, *Japanese Religion: Unity and Diversity* (Belmont, Calif.: Dickenson, 1969), 33. My summary, but such a phrase is often attributed to Prince Shotoku (?–622) who first authorized the three religions, i.e., Buddhism, Shinto, and Confucianism, for national worship.

11. According to Kenji Ueda, a voluntary liaison organization, the Association of Shinto Shrine founded in 1946, tried to formulate their creed and published a booklet in 1956. This shows that Shinto has not accepted a catechism for very long. See Ueada's "Shinto," in *Japanese Religion*, ed. Ichiro Hori et al. (Tokyo: Kodansha, 1972), 33.

12. Fortunately Jean Herbert published an excellent English work in 1967. I mainly depend on its arguments when I refer to Shinto.

13. Plumwood, *Feminism and the Mastery of Nature*, 47.

14. Plumwood, *Feminism and the Mastery of Nature*, 43.

15. Warren, "Feminism and Ecology: Making Connections," 6.

16. Val Plumwood, "Ecofeminism: An Overview and Discussion of Positions and Arguments," supplement to *Australian Journal of Philosophy*, (1986), 134; and Plumwood, *Feminism and the Mastery of Nature*, 44.

17. Plumwood, *Feminism and the Mastery of Nature*, 121.

18. Plumwood, *Feminism and the Mastery of Nature*, 123.

19. This argument also relates to her critique of deep ecology's concepts of self. In particular she criticizes the expanded self as "an enlargement and an expansion of

egoism" and the transcended or transpersonal self as "the conquest of personal ego." Also see Plumwood, *Feminism and the Mastery of Nature*, chapter 7.

20. Plumwood, *Feminism and the Mastery of Nature*, 126.

21. Plumwood, *Feminism and the Mastery of Nature*, 126.

22. Plumwood, *Feminism and the Mastery of Nature*, 128.

23. Plumwood, *Feminism and the Mastery of Nature*, 131.

24. Plumwood, *Feminism and the Mastery of Nature*, 133.

25. Jean Herbert, *Shinto: At the Fountain-Head of Japan* (London: Allen and Unwin, 1967), 234.

26. Herbert, *Shinto: At the Fountain-Head of Japan*, 235.

27. Herbert, *Shinto: At the Fountain-Head of Japan*, 252.

28. Herbert, *Shinto: At the Fountain-Head of Japan*, 256.

29. In the main body, Izanami is called "his younger sister." This is a literal translation. But it does not mean relatives. Both Izanagi and Izanami appear and are matured by themselves and, in general, this phrase means *wife* in ancient Japan (Herbert, *Shinto: At the Fountain-Head of Japan*, 248).

30. This relationship between washing the body and death is related to Shinto's crucial belief in purification.

31. She was born when Izanagi cleansed his left eye. Interestingly, the Moon "God" was born next when he cleansed his right eye. And in those days, left was always seen to be superior to right.

32. This is because our emperor issued his "Human Declaration" (not Human Rights Declaration!) after World War II.

33. Herbert, *Shinto: At the Fountain-Head of Japan*, 61.

34. Herbert, *Shinto: At the Fountain-Head of Japan*, 61–62.

35. Herbert, *Shinto: At the Fountain-Head of Japan*, 61–62.

36. Plumwood, *Feminism and the Mastery of Nature*, 155.

37. Plumwood, *Feminism and the Mastery of Nature*, 156.

38. Plumwood, *Feminism and the Mastery of Nature*, 159.

39. Augustin Berque, "The Sense of Nature and Its Relation to Space in Japan," in *Interpreting Japanese Society: Anthropological Approaches*, ed. Joy Hendry and Jonathan Webber (Oxford: Jaso, 1986), 101–102.

40. Chie Nakane, *Tekiou no jyouken* [Requirements of adaptation] (Tokyo: Kodansha, 1972), 138.

41. Earl Miner, Hiroko Odagiri, and Robert Morrell, *The Princeton Companion to Classical Japanese Literature* (Princeton, N.J.: Princeton University Press, 1985), 23.

42. Deane Curtin, "Dogen, Deep Ecology, and Ecological Self," *Environmental Ethics* 16, no. 2 (1994).

43. Curtin, "Dogen, Deep Ecology, and Ecological Self," 201.

44. Curtin, "Dogen, Deep Ecology, and Ecological Self," 201.

45. Curtin, "Dogen, Deep Ecology, and Ecological Self," 201.

46. Curtin, "Dogen, Deep Ecology, and Ecological Self," 198.

47. See Aiko Ogoshi and Junko Minamoto, *Kaitaisuru Bukkyou* [Anatomy of Buddhism] (Tokyo: Daitousyuppan, 1994), especially 177.

48. See also Jim Cheney, "Eco-Feminism and Deep Ecology," *Environmental Ethics* 9, 2 (1987): 120–29.

49. Warren, "The Power and the Promise of Ecological Feminism," 24–29.

50. Warren, "The Power and the Promise of Ecological Feminism," 33.

51. Plumwood, *Feminism and the Mastery of Nature*, 184–85.

52. Warren, "The Power and the Promise of Ecological Feminism," 31; and Plumwood, *Feminism and the Mastery of Nature*, 183.

53. Carol Gilligan, "Moral Orientation and Moral Development," in *Women and Moral Theory*, ed. Eva Feder Kittay and Diana Meyers (Lanham, Md.: Rowman & Littlefield, 1987), 22. My emphasis.

54. Gilligan, "Moral Orientation and Moral Development," 23.

55. Gilligan, "Moral Orientation and Moral Development," 26.

56. Herbert, *Shinto: At the Fountain-Head of Japan*, 69.

57. Important Kamis, including the Sun Goddess, are symbolically born when Izanagi washes his body after pollution. In Shinto, this "fact" is taken seriously.

58. Herbert, *Shinto: At the Fountain-Head of Japan*, 71.

59. Val Plumwood, "Nature, Self, and Gender: Feminism, Environmental Philosophy, and the Critique of Rationalism," in *Ecological Feminist Philosophies* (Bloomington: Indiana University Press, 1996), 172.

60. Plumwood, *Feminism and the Mastery of Nature*, 134–35.

61. Plumwood, *Feminism and the Mastery of Nature*, 135.

62. Masao Maruyama, "Rekisi ishiki no kosou" [The Basso Ostinato of historical view], in *Chyusei to hangyaku* [Loyalty and revolt] (Tokyo: Chikuma shobo, 1992), 334. In his article, Maruyama insisted that this historical view had been a "basso ostinat" throughout Japanese history and it had influenced every imported thought. The degree of influence of the worldviews identified in *Kojiki* and *Nihonsyoki* is debatable.

63. I translate *ikioi* as *energy*. But Maruyama states that in *Kojiki* and *Nihonsyoki*, *ikioi* means substantially the *virtu* that Machiavelli used as his key word. And in the Japanese case, it is not seen as energy because it is virtue, but it is revered as virtue since it has energy. See Maruyama, "Rekisi ishiki no kosou," 322.

64. Maruyama, "Rekisi ishiki no kosou," 339.

65. See, for instance: "Shinto insistently claims to be a religion of the 'middle-now,' the 'eternal present,' *naka-ima*. . . . Its main stress is on what should be done at the present instant, without much concern for what happened before or what will happen later, whether in this life or in after-life" (Herbert, *Shinto: At the Fountain-Head of Japan*, 32). Additionally, *naka-ima* means literally *between now*, or *here and now*.

66. Ueda, "Shinto," 40.

67. Roger King, "Toward an Ecological Ethics and Environment," in *Ecological Feminist Philosophies*, ed. Karen J. Warren (Bloomington: Indiana University Press, 1996), 82–96.

68. King, "Toward an Ecological Ethics and Environment," 91.

69. J. Baird Callicott, *Earth's Insights: A Survey of Ecological Ethics from the Mediterranean Basin to the Australian Outback* (Berkeley: University of California Press, 1994).

70. In this passage, he uses "purification" in Shinto's sense. But he also proposes a unique interpretation of this conception. According to him, "all the divinities of nature are born from the lower orifices of the feminine deity, whereas all the divinities related to culture . . . are born from the head of the male divinity," therefore, "The feminine deity represents the rotten, whereas the male divinity represents the pure." See Allan G. Grapard, "Nature and Culture in Japan," in *Deep Ecology*, ed. Michael Tobias (San Marcos, Calif.: Avant Books, 1984), 243. I cannot judge whether such a dualistic interpretation is right or not. For the moment, I would like to suggest that "he" is a "Western" scholar of Japanese "Buddhism," and that he thinks culture in Japan shows "male" purification of nature.

71. Callicott, *Earth's Insights*, 104. My emphasis.

72. Callicott, *Earth's Insights*, 104.

73. Callicott, *Earth's Insights*, 105. My emphasis.

74. Callicott is keener to discuss deep ecological elements in various religions in the world. He may have noticed this point *inversely*, because he recommends only Japanese *Buddhism* and *Zen* (not *Shinto*) for Japan to show the model of advanced and ecological society. Buddhism recommends a general and abstract love of nature, and Zen, especially the Zen taught by Suzuki to Westerners, is highly abstract. These two elements are common to deep ecology and it is this character that deconstructive ecofeminism condemns.

75. King, "Toward an Ecological Ethics and Environment," 92.

76. King, "Toward an Ecological Ethics and Environment," 92.

77. Max Weber, "The Profession and Vocation of Politics," in *Weber: Political Writings*, ed. Peter Lassman and trans. Ronald Speirs (Cambridge: Cambridge University Press, 1994), 359.

78. Weber, "The Profession and Vocation of Politics," 359.

79. Karen J. Warren, "Situated Universalism and Care-Sensitive Ethics" (paper presented to "Environmental Justice: Global Ethics for the 21st Century" conference, Melbourne, Australia, October 1997), 3.

80. Warren, "Situated Universalism and Care-Sensitive Ethics," 6.

81. Warren, "Feminism and Ecology: Making Connections," 3.

82. See the argument about the Shinto's two conceptions of Tama.

83. Warren, "Feminism and Ecology: Making Connections," 5.

84. See, for example, Warren, "Feminism and Ecology: Making Connections."

85. Plumwood, *Feminism and the Mastery of Nature*.

86. Plumwood, *Feminism and the Mastery of Nature*, 60.

87. The last category is death. A man can be polluted not only in this case but also through his wife's childbearing. Each category is also called "red pollution," "white pollution," and "black pollution." Interestingly, the colors red and white signify a happy event in Japan (for example, remember the Japanese national flag!). Perhaps this shows ambivalent feelings about these two matters. This is not true of black, which always signifies horrible things.

88. The more time passes, the more the perception of pollution concerning women has been stressed, and hence the pollution of women has been conceived essentialistically. However, there are disputes between Buddhism and Shinto about

the historical influence of this. I cannot judge this. I only propose the possibility of non-essentialism in this issue.

89. Christine J. Cuomo, "Unraveling the Problems in Ecofeminism," *Environmental Ethics* 14, 4 (1992): 354.

90. Cuomo, "Unraveling the Problems in Ecofeminism," 355. My emphasis.

91. Herbert, *Shinto: At the Fountain-Head of Japan*, 73.

92. Ogoshi and Minamoto, *Kaitaisuru Bukkyou* (Anatomy of Buddhism), 74.

93. In Shinto *Mikoto* is equal to *Kami*, although there is a debate about difference of nuances. And *Mochi* means *to hold*. Additionally, when a Japanese writes the word of biological life with a Chinese character, we use two ideograms, i.e., *Sei* (*living or life*) and *Mei* (Chinese pronunciation of *Mikoto*).

94. Ueda, "Shinto," 45.

95. Plumwood, *Feminism and the Mastery of Nature*, 154.

96. Plumwood, *Feminism and the Mastery of Nature*, 186.

11

Ecofeminists in the Greens

Greta Gaard

Introduction

Greta Gaard, an ecofeminist academic and activist, traces the role of ecofeminism within Green or ecological political parties in the United States. She documents the developments of ecofeminism, including approaches that characterize women's essential nature as caring and connected to the natural world, those that concentrate on radical social analyses, and others emphasizing spirituality. Her analysis of primary issues over ten years of "ecofeminism and the Greens" reveals the breadth of ecofeminist views. Greta also shows how new ideas, such as ecofeminism, make their way into political movements where they share some, but not all of the core values.

From Greta Gaard, "Ecofeminists in the Greens," in *Ecological Politics: Ecofeminists and the Greens* (Philadelphia: Temple University Press, 1998), 140–76. Copyright © 1998 by Temple University. All rights reserved. Reprinted by permission.

FROM PETRA KELLY IN WEST GERMANY to Ariel Salleh, ecofeminists have been central to the founding of the Green movement internationally. The United States is no exception. Indeed, the history of the U.S. Green movement bears the influence of numerous ecofeminists: Charlene Spretnak's organization of the founding committee and her contributions in drafting the Ten Key Values, Ynestra King's participation in the 1987 Amherst gathering, Marti Kheel's inclusion of animal rights in the life forms plank of the national Green platform, the statements from WomanEarth Feminist Peace Institute to the Greens' national program process at the 1989 Eugene gathering, Margo Adair's opening of many national gatherings with a guided meditation, Sharon (Shea) Howell's leadership in the "eco-city" project called Detroit Summer, Chaia Heller's efforts in the founding and Laura Schere's in the development of the Youth Greens, the many ecofeminists who contributed to the development of the Left Greens (Janet Biehl, Stephanie Lahar, Cora Roelofs, Joan Roelofs, Catriona Sandilands), Dee Berry's years of service as the Greens' clearinghouse coordinator, Anne Goeke's efforts in founding the Gylany Greens, among many others. The work of ecofeminists in the Greens has involved not only ecofeminist projects but also the more fundamental projects of feminism itself: achieving equal rights and representation for women (a liberal feminist goal), revaluing many characteristics and behaviors traditionally devalued as "feminine" (a radical feminist goal), and working to eliminate racism and classism within the organization as well as within the larger society (a socialist feminist goal).

Not surprisingly, ecofeminists have often predicted that the success of the Green movement would depend on the ability of Greens to uproot sexism and patriarchal behaviors within the movement. "The liberation of women," Ynestra King observed, reflecting on the 1987 Green gathering in Amherst, "must be central to Green politics if [the Greens] are to survive and grow."[1] In her appraisal of the West German Greens' first four years, Charlene Spretnak observed, "One problem all wings of the party have in common is sexism," a problem that had already caused several Green women to "keep some distance from the patriarchal style of politics" in the Bundestag or to identify themselves first with other movements.[2] After the Greens were voted out of the Bundestag in December 1990, Petra Kelly openly lamented the "permanent state of ideological warfare" between *fundi* and *realo* factions, a battle that Charlene Spretnak, in her memorial to Kelly, called "macho 'hardball politics.'"[3] Drawing from the experiences of the West German Greens to guide a potential U.S. Green movement, Spretnak concluded that postpatriarchal politics in the Greens would work best at the local, grassroots level. The history of the U.S. Green movement, however, reveals a shift in focus from local, grassroots politics to state and presidential politics. It is my thesis that this shift corresponds directly to—and may even be the result of—an attenuation of ecofeminist and feminist presence and perspectives within the movement. To trace this shift, I begin with the foundations of the movement: the Ten Key Values, and the role of spiritual/cultural ecofeminism in their development.

At the founding meeting of the Committees of Correspondence (CoC) in August 1984, sixty-two people convened to set the direction for a Green movement in the United States. Over the course of a weekend together, they decided what type of organization to set up, what to name it, and where to locate an informational clearinghouse. But the most important decision they made involved articulating a Green philosophy and political vision. As with the founding of the Greens, there are at least two different "origin stories" for the Ten Key Values.

According to Charlene Spretnak, there was general agreement among those present at the founding meeting that "more than the four pillars [from the West German Greens] was needed; people felt those concepts needed to be expanded, made more specific, and fine tuned." During a brainstorming session, numerous suggestions were recorded on a flip chart and copied down on paper, and a scribe committee was charged with combining these suggestions into a statement of key values and circulating it among the participants. The scribe committee consisted of Eleanor LeCain and Charlene Spretnak in Berkeley and Mark Satin in Washington, D.C. As Spretnak recalls, "The only major addition our committee made to all the suggestions was the idea of presenting the subtopics under each key value as a question for discussion, in order to invite participation in our grassroots organizing. That idea and others came from Mark Satin; Eleanor and I agreed, and so did everyone who had been at the founding conference and then received our draft in the mail."[4] Founding participants suggested fine tunings that were incorporated into the draft before it was printed in grassroots publications around the country.

A somewhat different story is told by Howie Hawkins, who also attended the founding meeting and later withdrew his support from the draft of the Ten Key Values.

Political philosophy was discussed indirectly in the acceptance of a draft of Ten Key Values as an initial discussion paper that would generally indicate the political direction of the CoC. Charlene Spretnak and Mark Satin were the principal writers who ran their drafts past the meeting periodically over the weekend. It was basically an expansion of what had been the West German Greens' initial basis of unity, the "four pillars" of Green politics—*ecology, nonviolence, social responsibility,* and *grassroots democracy.* Ecology was changed to "ecological wisdom," reflecting the spiritual and mystical bent of many present. Social responsibility was changed to "personal and social responsibility" to reflect the New Age emphasis on personal transformation. Added were decentralization, community-based economics, postpatriarchal values, respect for diversity, global responsibility, and future focus. Postpatriarchal values was a euphemism for feminism and respect for diversity for racial equality.[5]

What difference would it make whether the Ten Key Values emerged from brainstorming sessions with sixty-two participants, with LeCain, Spretnak, and Satin serving as scribes, or whether Spretnak and Satin brought to the meeting a Ten Key Values draft for commentary, incorporating the comments they received and circulating the new draft by mail for approval? Why were the terms "community-based economics," "postpatriarchal values," and "respect for diversity" chosen rather than the terms "anticapitalism," "feminism," and "antiracism"? In both narratives, Spretnak and Satin played significant roles, and the philosophies they represented—spiritual/cultural ecofeminism and New Age thought—have left their influence on the Greens' ideological foundation.

Spretnak had edited *The Politics of Women's Spirituality* and would soon author *The Spiritual Dimension of Green Politics,* and Satin was the author of *New Age Politics* and would become publisher of the New Age newsletter *New Options.*[6] Spretnak's position is further revealed in the following words: "Ecofeminism grew out of radical, or cultural, feminism (rather than from liberal feminism or socialist feminism), which holds that identifying the dynamics—largely fear and resentment—behind the dominance of male over female is the key to comprehending every expression of patriarchal culture with its hierarchical, militaristic, mechanistic, industrialist forms."[7] Spretnak's and Satin's books drew upon radical and cultural feminist critiques of women's oppression. These critiques, which had emerged from women's disillusionment and subsequent separation from the New Left of the sixties, argued that men's domination of women under patriarchy was the root cause of all other oppressions.

The sixty-two founding Greens may have chosen the term "community-based economics" over "anticapitalism" because Spretnak and Satin rejected leftist critiques (both cultural feminism and New Age thought are antileftist). They preferred the slogan from the West German Greens, which had graced the cover of Spretnak and Capra's *Green Politics:* "We are neither Left nor Right; we are in front." Satin's antileftism permeates his *New Age Politics,* whose title concept he defines as moving "beyond liberalism," "beyond Marxism," and "beyond the Anarchist alternative."[8] "Postpatriarchal values" may have been chosen over "feminism" for at least two reasons: First, the conservative backlash against feminism was already underway, and many Greens wanted to avoid any negative associations that the term "feminism" might invoke. Second, the term "postpatriarchal values" is closely tied to ideas about a paradigm shift, a concept that was very popular in the early eighties. New Age writers and intellectuals observed that behaviors and traits of an old paradigm were falling from favor and a new paradigm was being discovered. At the same time, radical and cultural feminists discussed many of these same concepts in terms of patriarchal versus matrifocal or woman-centered cultures. The old paradigm (patriarchy) was associated with hierarchy, competition, militarism, top-down leadership, while the new paradigm (women's culture) was associated with networks, cooperation, nonviolence, bottom-up decision making, to name a few of the behaviors linked to each.

The most salient problem with the Ten Key Values is that they do not offer a critique of the way that race and class have influenced the development of technologies, economies, or social and political systems that are ecologically destructive—and they do not specify the need to eliminate capitalism or racism as integral to an ecological agenda. Significantly, neither New Age thought nor cultural feminism offers an analysis of capitalism or racism. For New Age thought, this absence is explained by its emphasis on the primary importance of personal transformation. For cultural feminism, the absence is explained by its radical feminist origins. Radical feminism holds that the oppression of women and the associated devaluation of the "feminine" under patriarchy is the root cause of all other oppressions; thus going to the root, by ending women's oppression and valuing all those things associated with the "feminine," would of necessity end all other oppressions as well.[9] Janet Biehl, a cofounder of the Left Green Network (LGN) and a social ecofeminist, has argued, "Dropping capitalism and statism from direct consideration in feminist theory renders feminism nonrevolutionary" and subject to co-optation; to avoid this problem, she has urged, "it is high time that ecofeminists challenged the notion of 'primary oppression' and thereby rekindled discussion of the relationship of feminism to the left."[10] Hence, while the analytical strength of cultural ecofeminism and the Ten Key Values lies in their emphasis on personal and cultural transformation in achieving social and ecological justice, this strength is also their limitation. Social ecofeminists felt that by excluding specific critiques of other systems of oppression, such as capitalism and racism, the Ten Key Values set up a Green movement that was ripe for takeover by a more conservative interpretation. Social feminists saw some of these conservative elements manifest in deep ecology and they promoted a strong debate on these issues among Greens.

At the first national Green gathering in Amherst in 1987, the social ecology/deep ecology debate received much attention, and in the two years between the Amherst and Eugene gatherings, while academic ecofeminists were debating deep ecologists in the pages of *Environmental Ethics*, social ecofeminists in the Green movement were resisting overtures from deep ecologists and bioregionalists alike. In 1987, the _Nation_ published Kirkpatrick Sale's "Ecofeminism—A New Perspective" and Ynestra King's social feminist response, "What Is Ecofeminism?" In his essay, Sale described ecofeminism as a "hybrid" of feminism and ecological politics (free of influence from leftist politics), an "amalgam" that articulates women's new understanding, based on experiences of working for political transformation in the environmental and Green movements, that "the problems are of culture and values more than politics and laws." He portrayed ecofeminism as combining "that 'scientific' thought so often the province of men and the intuitive experience of subjugation and exploitation known to women" and suggested four "hallmarks of present ecofeminist thought": goddess cultures, earth-based spirituality, bioregionalism, and deep ecology.[11]

Sale cited approvingly the strategy of combining gendered attributes to create a holistic analysis, a strategy used by both cultural ecofeminists and deep ecologists. Cultural ecofeminists and deep ecologists share a strategy of reversing valuations in the classic culture (man)/nature (woman) dualism: deep ecologists urge humans to subordinate themselves to nature (biocentrism), and cultural ecofeminists celebrate women's connections to nature and many traditionally feminine characteristics. Sale correctly perceived cultural ecofeminism's emphasis on personal and cultural transformation as more significant than—or, of necessity prior to—political, legislative, and economic transformation. Moreover, his focus on goddess cultures and earth-based spirituality as the salient features of ecofeminism were accurate, although this branch has many other political tendencies as well; cultural ecofeminism is distinguished by its emphasis on spirituality not because that is its only emphasis but because no other branch of ecofeminism has so fully recognized the importance of spirituality. The fact that Sale cited bioregionalism and deep ecology as features of ecofeminism can be attributed to Judith Plant's involvement in the bioregional movement and Charlene Spretnak's frequent pairing of ecofeminists and deep ecologists in a way that suggested this was a coherent and comfortable grouping.[12]

Ynestra King's response to Kirkpatrick Sale offers a clear articulation of social ecofeminism's significant differences from cultural ecofeminism, bioregionalism, and deep ecology:

> Although most of us come from the left and maintain a commitment to the leftist projects of human liberation, historical analysis and an opposition to capitalism, we share the social anarchist critique of the economism, workerism and authoritarianism of a myopic socialism that has not challenged the domination of nonhuman nature or taken ecology seriously. We both extend and critique the socialist tradition, sharing with socialist feminist theory an analysis of patriarchy as independent of capitalism, and with cultural feminism an appreciation of

traditional women's life and work. Also, taking ecology seriously has meant that we can't opt for a piece of a rotten, carcinogenic pie (like liberal feminists) or set up the woman/nature connection as the enemy of feminism. Connecting women to nature need not acquiesce to biological determinism (the legitimate fear of socialist feminists) if nature is understood as a realm of potential freedom for human beings—both women and men—who act in human history as part of the natural history of the planet, in which human intentionality and potentiality are an affirmed part of nature.[13]

King's affirmation of the leftist, anarchist, and feminist contributions to ecofeminism stood in clear contrast to the cultural ecofeminism described by Sale. In addition, King rejected bioregionalism for its failure to provide an analysis of human oppression: small, ecologically sensitive communities can still foster racism, sexism, and homophobia, all of which are unacceptable from an ecofeminist perspective. Deep ecology also ignores entrenched structures of human oppression, focusing exclusively on the "laws of nature"—as defined by deep ecologists.

King's critique of deep ecology was extensive. First, King (drawing on Marti Kheel's critique) cited "the pro-hunting stance of the deep ecologists" and the defense of "privileged white males of the developed Western countries . . . [going out to] kill something *to realize their identities* as natural beings" as an example of deep ecology's inherently masculinist bias. She described deep ecology's rejection of distinct human selves as fundamentally antithetical to the feminist project of reconstituting women's identities and perspectives. She thought that deep ecology's rejection of anthropocentrism for biocentrism was a simple inversion and cited the commitment to ecological humanism as the crucial difference between ecofeminism and deep ecology. Finally, King rejects the Malthusian wing of deep ecology as deeply misogynist, racist, and homophobic, "with no analysis of U.S. imperialism, corporate capitalism, the debt of the Third World to the First and the enforced growing of cash crops to pay our banks as the causes of famine in the Third World and enormous suffering in Central America." Population will not be controlled until women have economic and social power and the radical "social, racial, and economic inequities around the world" are addressed. King's conclusion that social movements must strive first "to end the domination of human over human in order to end the domination of people over nonhuman nature" is a succinct articulation of social ecology's core principle,[14] and it stands in direct opposition to both deep ecology and cultural ecofeminism.

Sale's description of ecofeminism can be seen as one example of a larger struggle within the Green movement between the perspective that ecofeminism is a subsidiary of the Green movement and the perspective that ecofeminism is a distinct movement. Although neither King nor Sale explicitly remarks on this phenomenon, Sale's definition—and King's emphatic redefinition—highlight the tension between merger and autonomy. As Janet Biehl observed, "Women know from long experience that when they are asked to become 'one' with a man, as in marriage, that 'one' is usually the man. Ecofeminists should be equally suspicious of this 'ecological' oneness." The invitation to merge was coming from men in the arena of ecological politics and philosophy, and in both cases it meant erasure of ecofeminist identities, philosophies, and perspectives. In 1988, Biehl pointedly critiqued the sexism in deep ecology's "advances" toward ecofeminism, concluding that "ecofeminists have nothing to gain in such an embrace."[15] Biehl's critique focused on four specific areas in which deep ecology puts women in a double bind. First, Biehl observed that deep ecology's identification of anthropocentrism as the central problem of modern Western industrialized culture treats humans as an undifferentiated whole. At the same time, however, deep ecologists affirmed and even sought to emulate women's position as "closer to nature," a celebration they shared with cultural ecofeminists. But as Biehl pointed out, by holding women equally responsible with men for an anthropocentric position that has devalued and destroyed nature, and simultaneously perceiving women as "closer to nature," deep ecology leaves women with nowhere to stand.

The deep ecological self, based on a connectedness that erases difference, is both celebrated and sought after; moreover, this connected self is portrayed as a preexisting feature of women's psychological makeup, one that is to be emulated by men. Biehl cited feminist psychological research showing how women's soft or permeable ego boundaries have prevented women from pursuing their own rights and interests. But, Biehl explained, women's search for selfhood is "the revolutionary heart of the feminist and ecofeminist movements." Being told to "think like a mountain" when we are now beginning to find our own consciousness as women is "a slap in the face." This is deep ecology's second double bind for women: It seeks to deny difference at the same time it seeks to appropriate an aspect of feminine psychology. Deep ecology's fetishization of wilderness perpetuates a culture/nature dualism that is deeply ethnocentric. As Biehl explained, most Native American cultures have no word for "wilderness" because there is no hyperseparation between humans and nature in preindustrial, ecologically sustainable cultures. For ecofeminists, deep ecology's conception of wilderness as a "sacred space" in which to heal from the alienation of mechanized society becomes particularly relevant in a context where women and nature are portrayed as wild and chaotic, a portrayal that has been used to justify the domination of both. Thus conceptualized as healers, women and nature are supposed to play the nurse to wounded male deep ecologists. Here Biehl made one of her most radical assertions, one that ecofeminists are only beginning to pick up on ten years later: "Women are not 'chaotic' but rational," Biehl argued, "and nature, too, is not 'chaotic' but rather follows a logic of development toward increasing complexity and subjectivity, replete with differences, individual variations, and the slow formation of selfhood."[16] Although it would be unlike a social ecologist to argue that nature has *reason* (the defining characteristic of humans, according to social ecologists), Biehl came fairly close to it by arguing for nature's "logic." Deep ecology thus denies reason to women and logic to wilderness at the same time it requires women and wilderness to provide the space for healing so men can restore their own wholeness. Therefore, the wholeness of deep ecology depends on the denial of wholeness to women and wilderness.

Finally, deep ecology's focus on the problem of overpopulation and its solution of fertility programs contains an inherent contradiction: on the one hand, deep ecologists deplore the progress of industrial society; on the other hand, they demand fertility programs, which have been made possible largely through the progress of industrial society. Moreover, their solution overlooks the real solution to the population problem, which is feminism itself. But this solution would blow apart all the other features of deep ecology, with women thinking like mountains, being wild and chaotic, and being closer to nature. In sum, Biehl urged ecofeminists to resist "deep ecological self-oblivion."[17]

The warning was an important one. Some male deep ecologists, critical of ecofeminism, were fond of saying that deep ecology had already articulated the most valuable insights of ecofeminism: a sense of self as interconnected with nature, earth-based spirituality, and a rejection of Western culture's over-rationalization, mechanization, and isolation. In these debates with deep ecologists and with Greens, social ecofeminists struggled to retain a distinct perspective and a distinct movement. At the same time, ecofeminists were working to bring an ecofeminist perspective into the Green movement through their activism—through their participation in building Green locals, attending national gatherings, and developing the national Green platform.

Ecofeminists of all stripes worked to build the Green Committees of Correspondence (GCoC) between 1989 and 1991, the years of annual national Green gatherings at Eugene, Estes Park, and Elkins. This period arguably marked the height of ecofeminist involvement in the GCoC; they were also the years of the nearly sacrosanct status of liberal and cultural feminisms in influencing certain ways of proceeding. Liberal feminism's influence could be seen in the fact that gender balance was required in all administrative bodies and in the processes of every meeting: regions had to send two representatives, a man and a woman, and discussions were frequently organized by placing potential speakers in a "stack" and alternating speakers between men and women. Cultural feminism's influence could be found in the fact that decisions were made not by voting but by consensus, a process that was seen as much more feminist. Gatherings were often opened with a guided meditation, and "women's ways of knowing"—that is, intuition, feelings, relationships, spirituality—were given special recognition (in theory, if not always in actuality). Moreover, collaborative writing was common; in fact it was the process used to develop the national Green platform, and has often been claimed as a feminist process because of its inclusivity and its reliance on frequent and effective communications among participants. Yet, in spite of these attempts to bring a feminist perspective into the Greens, ecofeminists were never able to gain much of a standing within the movement. Now I examine three instances wherein feminist and ecofeminist ideas were made available but failed to generate support: consensus decision making, the proposed statements for the Green national platform submitted by Woman-Earth and by the Life Forms Working Group, and antiracism training and projects.

"Voting is violence," as one Green became famous for remarking, based on the critique that traditional voting excludes and disempowers the minority for the sake of the majority. In contrast, consensus decision making is supposed to give each person power equal to the sum of everyone else. To facilitate consensus decision making at the national Green gatherings, organizers called on Caroline Estes, a feminist bioregionalist and cofounder of the intentional community Alpha Farm. As Estes explained, *Robert's Rules of Order* were crafted by Colonel Robert during the gold rush frenzy of San Francisco in 1867.[18] Surely, she argued, this procedure for decision making retains the militaristic, competitive, and acquisitive mentality of the context in which it was crafted. Consensus decision making, on the other hand, came out of the Quaker tradition and was designed to build community through cooperation, attentive listening, and the belief that although every person has a part of the truth, no single individual has it all. Apparently taking "feminine" to be the equivalent of "feminist" (a hallmark of cultural feminism), Estes still believes consensus is indeed a feminist process because "it's more inclusive, more compassionate; it requires paying attention to everybody, and it brings about the same kind of caring as women do in mothering."[19] Many people believed this "nonviolent" process of decision making would be well suited to the Greens.

Unfortunately, it wasn't. Instead, the consensus process was blamed for the near paralysis and inefficacy of the Interregional Committee, the national coordinating body of the Green movement. Left Greens and Youth Greens alike were consistently critical of the consensus process, because, when one person was capable of blocking a decision that the majority was ready to carry out, consensus effectively allowed minority rule. One Left Green pointed out that "to insist that nothing happens in the name of the Greens until all agree is, in practice, to insist that nothing happens."[20] Several years after she retired from facilitating the national gatherings, Caroline Estes reflected on the reasons that consensus had failed in the Greens. "They weren't trained in consensus," said Estes, "and the political agendas were too strong. If you come in with your mind made up, consensus won't work."[21] In sum, the Greens made the error of mistaking a feminine or cultural feminist process as a one-size-fits-all feminist process, but feminism is characterized by its attention to context, not its rigid adherence to rules or methods. A genuinely feminist process would have involved Greens in assessing the movement as it really was (not how they wished it might be) and then adopting or developing a process that suited their unique needs. The consensus process was eventually replaced by falling back to a simple majority vote when consensus could not be achieved.

In other instances, Greens could not be moved to accept ecofeminist insights, particularly when it came to the platform itself. At the 1989 national gathering in Eugene, for example, six statements were submitted to the platform process on behalf of the WomanEarth Feminist Peace Institute. Included were statements on ecofeminism written by Ynestra King, Irene Diamond, and Charlene Spretnak; on spirituality written by Spretnak and Starhawk; on racism titled "Toward Diversity" written by Margo Adair; on bearing and caring for children written by Irene Diamond; on animal rights written by Marti Kheel; and a statement on nonviolence by Spretnak. Each of the statements articulated a feminist understanding of the personal/political connection—a perspective that distinguishes ecofeminism from both social ecology (which critiques social structures rather than cultural values) and deep ecology (which emphasizes personal transformation but does not critique social or economic structures other than to defend wilderness from their operation). In the statement defining ecofeminism, Ynestra King explained that the ecological crisis is inseparable from the many social crises, and Irene Diamond discussed ecofeminism's critique of dualisms, positioning women as part of both nature and culture. Charlene Spretnak described the ecofeminist contributions to the Green movement as including "not only analysis, theory-building, and policy recommendations, but also an emphasis on feminist process, without which we feel the Green movement will never reach its potential." They concluded by emphasizing an ethic of care.[22]

In the statement on spirituality, Spretnak and Starhawk explained the necessity for personal as well as political transformation: "We value all spiritual practices that help cleanse the mind of ill will so that love and compassion can flourish. Dominance and exploitation will not end as long as hatred, cruelty, and indifference drive so much of human action." Margo Adair made the same personal/political connection in the statement on diversity, which she began by observing, "Racist socialization is the glue that keeps institutional racism in place." In the statement on bearing and caring for children, Irene Diamond connected the private with the public in her criticism of reproductive technologies, her defense of women's right to choice as a necessary component of women's participation in the politics of the public sphere, and her incisive analysis of Western culture's fear of death. Marti Kheel's twelve-point plank on animal liberation positioned nonhuman animals within the Green values of nonviolence and ecological wisdom and cited from the GCoC brochure: "There is no solution to the ecological crisis that fails to uproot human domination in all its forms." And in the statement on nonviolence, Spretnak connected militarism and poverty, domestic violence and cruelty to animals, the economic enslavement of Third World nations and racism in the United States. All the WomanEarth statements emphasize the theme of interconnection.

None of the statements survived in the final version of the Greens' national platform. Organizers shifted the Eugene conference from a decision-making body to a gathering focused on learning and discussion, with the understanding that working groups would develop specific statements through a clearly explained process over the next year and would present those statements for a vote at the 1990 gathering in Estes Park. It was shortly after the Eugene gathering that John Rensenbrink recruited political science professor Christa Slaton to oversee the collaborative writing process and persuaded Margo Adair, editor of *Green Letter*, to assist in the effort by publicizing regular activity reports and platform drafts. The point here is that even with a feminist writing process, along with feminist (Slaton) and ecofeminist (Adair) coordinators, the Green national platform still did not put forward the WomanEarth statements. The ecofeminist statements and perspectives were directed to other issues or platform planks, where they were later subsumed or voted down.

The members of the Life Forms Working Group faced their own set of challenges in getting their plank adopted by the larger Green movement. The group was ably facilitated by a number of experienced activists: Marti Kheel, whose work on animal ecofeminism was becoming widely known, was the spokesperson for the group. Charles Dews, a Green who had written widely about the need for the animal liberation and Green movements to work in coalition, served as co-coordinator. And Connie Salamone, a woman who might well be known as the "mother" of the animal liberation/feminist connection, worked on the planks while participating in a New York City Greens local. Together they crafted statements using each of the Ten Key Values to demonstrate that speciesism was antithetical to a Green vision. Then they put forth a plank addressing the retainment of origins in natural gene pools (addressing plant and animal concerns as well as genetic engineering), the preservation and sustainability of ecosystems, and the quality of life for animals and plants alike.

In April 1990, just months before the gathering in Estes Park, the Life Forms Working Group received a critique of its statements from the Rocky Mountain Greens. In each case the Rocky Mountain Greens' objections were aimed at the ecofeminist perspective in the statements, not the specific recommendations. They argued that the value of nonviolence may not be extended to nonhuman animals, not because of any inferiority of other species but because "life on the planet Earth is by nature violent." They also objected to the application of postpatriarchal values to other species, claiming that the attribution of all oppression, including speciesism, to "patriarchal social and economic structures" was a form of "male-bashing." On applying the value of community-based economics to nonhuman species, they rejected the analysis that human and nonhuman species were exploited by multinational corporations, arguing that "this statement ignores individual responsibility." And the Rocky Mountain Greens vowed to block consensus unless the Life Forms Working Group chose to rewrite its statements. The critique reveals that a basic resistance to an ecofeminist perspective existed, but, despite their objections, most of the recommendations from the Life Forms Working Group survived the platform ratification process at Estes Park and were included in the Green national platform. Nevertheless, the application of the Ten Key Values to nonhuman species did not go forward, nor did the people working in the Life Forms Group. Marti Kheel, for example, was exhausted by the process of battling with people unwilling to listen to alternative viewpoints.

Finally, Shea Howell and Margo Adair were two ecofeminists whose work within the Greens was largely underappreciated. As ecofeminists, both women had attended the West Coast meetings of WomanEarth II in 1989, and they were among those white women who volunteered to leave the meeting in order to allow those who remained to achieve racial parity. Not only had Adair been selected to co-coordinate WomanEarth had it survived; she also co-facilitated the Greens' platform writing process during 1989–1990, edited *Green Letter,* and brought an ecofeminist perspective to the Green gatherings through her guided meditation exercises that opened many meetings and her frequent workshops on unlearning racism. Together Howell and Adair operated Tools for Change, offering training programs to groups and organizations on uprooting relations of domination and on alliance building, with a particular focus on race and class. The leaflet "Toward Healing All Our Relations," abstracted from their pamphlet *The Subjective Side of Politics,* was often the only and always the most-cited resource in the Green movement for understanding racism, sexism, classism, and other forms of domination. Moreover, Detroit Summer was coordinated by Shea Howell.

The idea for Detroit Summer was first created by Detroit urban activists James and Grace Lee Boggs as a way to address community problems through direct grassroots activism. Sharon Howell, Roberto Mendoza, and the Detroit Greens immediately became part of the effort, planning projects and contacting community activists. Before the project was introduced to the Greens in 1991, the people of Detroit had already formed "112 neighborhood organizations, thirty neighborhood small business associations, three hundred cooperatives in housing, food, day-care and worker collectives."[23] Recognizing the efforts of Detroit activists as a model of Green theory in action, Greens adopted a Detroit Summer project as part of a tripartite Green Action Plan for 1992. The agreement was to send students and other Green volunteers to Detroit to assist in the efforts that were already under way. But as Howell recalls, the Greens' national membership never followed through on this project, which was intended to build multicultural, intergenerational alliances. In spite of the good efforts made by Howell and Adair, the project of unlearning sexism and racism—much less building multicultural alliances—went largely undone.

I turn now to the involvement of social ecofeminists in the Left Greens and the Youth Greens. Although the term "social ecofeminism" hadn't been coined, it was a principle shared by the Left and Youth Greens from the start. According to Chaia Heller, Murray Bookchin first began using the term "ecofeminism" in 1976 as he mapped out the curriculum for the Institute for Social Ecology (ISE).[24] In 1980, Ynestra King and other feminist activists organized the Conference on Women and Life on Earth: Ecofeminism in the '80s, a gathering inspired by King's study and teaching at ISE as well as her commitment to developing a feminist peace politics. But it was not until the mid-eighties that Chaia Heller, who had also been studying and teaching at the ISE, began using the term "social ecofeminism" as a way of acknowledging the influence of social ecology as well as distinguishing this particular version of ecofeminism. Although she frequently lectured on social ecofeminism, and thus her ideas were widely available among Left Greens, Youth Greens, and others at the ISE, Heller did not write up her notes in essay form until 1990.[25] But the term became explicit by October 1988 when Janet Biehl used it in her essay, "What Is Social Ecofeminism?" It was adopted as a founding principle of both the Left Greens and the Youth Greens less than a year later.

In April 1989, at their founding conference in Ames, Iowa, members of the Left Green Network offered a succinct definition of social ecofeminism as one of their fourteen founding principles:

Left Greens are committed to the liberation of women, to their basic reproductive rights as well as their full participation in all realms of social life. We believe in a social ecofeminism that seeks to understand and uproot the social origins of patricentric structures of domination. Unlike other ecofeminisms that accept patriarchal myths and cultural definitions of women as more "natural" than men and as existing outside culture, social ecofeminism regards women as cultural beings, as well as biological beings, and seeks to understand and change the social *realities* of the relationships between women, men, the political realm, the domestic realm, and all of these to nature.

Students at Antioch College, inspired by presentations and discussions with Chaia Heller, John Clark, and Howie Hawkins, formed the Youth Greens during their weekend conference on Green politics in May 1989. Writing position papers on each of six core principles, the Youth Greens were perhaps even more explicit than the Left Greens in their analysis of women's economic, individual, and cultural oppression: "Social Eco-feminism in the Youth Greens recognizes capitalism's inherent oppression of women and others. We critically analyze the relationships between power, gender, and social structure, and the relationships among all forms of oppression. We actively pursue egalitarian principles through affinity groups, active dialogues between men and women, and corrective measures on a personal and group level."[26] At their third national gathering in July 1990, Youth Greens developed a two-page ecofeminist statement defining ecofeminism as an analysis of the interconnection of all forms of domination. From the perspective of social ecofeminism, Youth Greens explicitly identified the role of capitalism in perpetuating women's oppression. Both the Left Greens and the Youth Greens were careful to define social ecofeminism as a theory that "recognizes the historical connections between the domination of women and the degradation of 'nature'" but that rejects the dualism of woman/nature as opposed to man/culture. According to the Youth Green statement, "Ecofeminists are laying the myth of biological determinism to rest."

Neither the Youth nor the Left Greens accepted social ecofeminism uncritically, and feminists in both groups had to struggle with sexism. At the founding of the Youth Greens at Antioch, for example, Chaia Heller was "met with a fair amount of animosity" from young men who felt her speeches were "aggressively lesbian" and accused her of "recruiting for the eco-feminist movement."[27] Women who formed a separate women's caucus as a space to discuss Green theory issued a statement to other conference participants: "it is obvious that the prevailing social oppressions of our society have not been transcended by the youth green movement" and emphasizing that "the values and contributions of [a feminist] caucus are necessary for the success of the youth green movement and we expect full support." In spite of these early difficulties, several ecofeminists described both the Left Greens and the Youth Greens as some of the most politically astute activists with whom they had ever worked.

In terms of direct action, the clearest example of the convergence of ecofeminism and Green politics was manifested in the Earth Day Wall Street Action. As models for their action, members of the organizing committee drew on the examples of two previous demonstrations: the Wall Street Action of October 29, 1979, organized by the antinuclear Clamshell Alliance on the fiftieth anniversary of the great stock market crash, and the Women's Pentagon Actions of November 1980 and 1981. In creating the *Earth Day Wall Street Action Hand-book,* organizers referred to the handbook from the Clamshell Alliance's Wall Street Action, adding explicitly ecofeminist essays by Chaia Heller, Margo Adair and Sharon Howell, Judi Bari and Ynestra King, and perhaps most important, a full reprint of the Unity Statement from the Women's Pentagon Actions. Finally, organizers modeled the day's events on the scenario developed for the Women's Pentagon Actions, which had four stages: mourning, rage, empowerment, and defiance.[28] For the Earth Day Wall Street Action, organizers sent mass mailings to all two hundred fifty Green locals, all locals and contact persons of the National Anti-Toxics Campaign, and hundreds of student groups, inviting them to submit two-hundred-word statements describing how their communities had suffered from out-of-control corporations. Like the Women's Pentagon Actions' ribbon of life, this Scroll of Anger and Mourning was intended for prominent display during the noon speak-out, when coalition members addressed demonstrators, stock exchange workers, and passersby. In theory and in practice, the Earth Day Wall Street Action of the Left Greens and the Youth Greens articulated an ecofeminist perspective.

Of course, ecofeminists within these direct action groups still had to combat sexism. At the Second Continental Conference of the Left Green Network (LGN) in July 1990, the women's caucus discussed the need to develop a feminist agenda for the LGN. Such an agenda would include changing the communication format of the conferences, which was characterized by a "polemical, repetitive oratorical style" and large-group plenaries "dominated by the rhetorically 'strongest' men in the room." As an alternative, the women's caucus suggested small-group discussions, circulating drafts of articles for discussion before the conferences, and assigning a "process watcher" to assist the facilitator in ensuring that everyone had a chance to participate.[29] Later that year at the Youth Green gathering, women confronted a similar problem in their gender caucuses. In both the Left Greens and the Youth Greens, ecofeminists recognized the value of working in coalition with a broader movement but emphasized the importance of keeping women's caucuses to maintain their identities as distinct from the group. Although feminism was implicit in an anarchist agenda—and social ecofeminism was explicit in the principles of Left and Youth Greens—"as long as there continue to be remnants of sexism in any organization," one Left Green woman explained, "feminists need the strength, clarity and definition that autonomous women's groups provide."[30]

It came as a surprise to many social ecofeminists, then, when Janet Biehl's new book, *Rethinking Ecofeminist Politics*,[31] advocated abandoning ecofeminism in favor of social ecology, a theory she believed would assuredly address the concerns of women. Although the Left Green organizing bulletin published a wholly uncritical review of Biehl in an issue just before the Third Continental Conference, at the conference itself Laura Schere and Kate Sandilands offered workshops on feminism that took Biehl's work in two different directions. Using Biehl's conclusions as a departure point, Schere warned against the Left Green tendency to collapse individual identities in the desire to create a unified Left. From Schere's perspective, ecofeminists were making crucial contributions to the understanding of domination; rejecting their insights would mean creating a "prematurely unified theory and practice that obscures real historical differences." Sandilands affirmed women's subjectivity, but with a different strategy. Criticizing the Left Green principle of social ecofeminism for its divisiveness, its "double-distancing from feminism," and its obsession with "defining what is wrong with all hitherto existing attempts to combine feminism and ecology," Sandilands advocated discarding the principle entirely and replacing it with "women's liberation." Apparently, a number of the women at the conference agreed with her, and the principle of social ecofeminism—along with any references to ecology—was dropped. In 1991, with the publication of Biehl's book and the Chicago conference of the Left Green Network, the brief popularity of social ecofeminism in the Greens was nearly over.[32]

Too late to have any real effect, the editors of the Left Green Network's discussion bulletin, *Regeneration*, sent out a call for papers to be submitted for a special issue on feminism. The call was circulated in fall 1991, after the fifth (and final) conference of the Youth Greens and the contentious national Green gathering in Elkins, which reorganized the Greens' national into the Greens/Green Party U.S.A. (G/GPUSA), subordinating the party to the movement in such a way that, just six months later, members of the Green Party Organizing Committee (GPOC) left the organization altogether and formed the Green Politics Network (GPN). The Left Green Network would have one more continental conference, in May 1992, before its membership numbers dropped. In short, the "movement" aspect of the Greens was waning. Although the call in *Regeneration* engendered several responses, overall the editors could only reflect, not resuscitate, the views and the presence of social ecofeminists.

In August 1991, a month after the Left Green Network's conference in Chicago, the Green Committees of Correspondence (GCoC) held their fourth national gathering in Elkins, West Virginia. The gathering has been discussed as a turning point in the Greens' national organization, marking as it did the last time party and movement activists worked together in a single national organization. Ecofeminism—or rather, the rhetoric of liberal and cultural feminisms, and cultural ecofeminism—played a significant role in the division of Green activists into two separate national organizations. According to the activists who formed the Green Politics Network (GPN), the blatant sexism of the Greens in the GCoC (combined with their own desire to form an electoral organization) motivated the Green Party Organizing Committee activists to withdraw from the organization. According to the women who remained in the newly reorganized Greens/Green Party U.S.A., the sexism they faced in the movement was no more and no less than women face in the larger society, and they charged the GPN activists with using the rhetoric of feminism to cloak a very patriarchal drive for political power. At the 1992 gathering that followed Elkins, many of these same women introduced a resolution to change one of the Ten Key Values from "postpatriarchal values" to "feminism." Thus, within the period of a year, the central articulations of the Green movement had been revised to emphasize "women's liberation" (LGN) or "feminism" (G/GPUSA), and during this time the Green electoral activists withdrew from the movement, in part because of its inability to transcend patriarchal behaviors. It is my thesis that, as with *Regeneration's* special issue on feminism, the changes in core values came too late and offered little more than lip service to solving a fundamental problem with sexism and patriarchal behaviors. I see a correlation between the disappearance of social ecofeminism and the unrestrained growth of patriarchal politics-as-usual in the Greens. Moreover, I find a disturbing compatibility between patriarchal politics and cultural ecofeminism.

Criticisms about patriarchal behaviors in the Greens' national organization began surfacing in 1991 and 1992. At the 1991 national gathering in Elkins, the preconference meetings of the Green Party Organizing Committee included the presence and voices of ecofeminists speaking on behalf of the GPOC. John Rensenbrink yielded a third of his allotted speaking time to Terri Williams of St. Louis, an ecofeminist who had split off from the Gateway Green Alliance in St. Louis because the "male-oriented, patriarchal Left Greens had taken over." Later, former national clearinghouse coordinator Dee Berry received extra time to express her ecofeminist analysis of the problem. According to Berry, the party/movement debates were exacerbated by "this patriarchal Left Green behavior," at which point another GPOC leader, Barbara Ann Rodgers-Hendricks of Florida, burst into tears.[33] With three women objecting to sexism, it seems there was surely some foundation to their claims.

Nor were they alone in their charges. Christa Slaton, the political scientist who had coordinated the national platform writing process from 1989 to

1990, left the national gathering at Estes Park, and the entire Green movement. In her review of the U.S. Green movement, Slaton cites several women who express frustration with "the dominance of males practicing politics as usual." Although many local and statewide Green groups had attempted to implement the value of postpatriarchy through liberal feminist strategies, still "males tended to dominate the process." According to Slaton, the problems reached a peak at the gathering in Elkins, after which "many members and former leaders—particularly women—broke away and formed a parallel organization called the Green Politics Network (GPN)."[34] Although she could not be induced to continue her participation in the Greens, no matter what the form, Slaton agreed to be one of thirteen signatories to the "Rationale for Launching a Green Politics Network."

There was a buildup to that statement, of course, involving first the resignations of Barbara Ann Rodgers-Hendricks and Dee Berry from the Green Council, along with various private communications among GPOC members. After her initial letter of resignation, Berry followed up publicly in December 1991 with another letter to the Green Council and members of the G/GPUSA. In this letter, Berry said: "There is a creative female energy in both men and women at the very depths of our beings that is a strong force for life. This force, like life itself, is sexual, spiritual, messy, ecstatic, unpredictable and represents death as well as life. Patriarchy has always been afraid of this energy because it cannot control it or reduce it to fit its neat intellectual theories. So the reaction has been to violently oppress it." To defend women from being overwhelmed in the struggle with patriarchal forces, Berry invokes the strength of the Green warrior:

> We must integrate our female energy with the power of the Green warrior, the positive male force of both women and men. Unlike the soldiers of patriarchy who get their power by destroying or putting down others, Green warriors get their strength from being who they are. Thus, to use the positive male force is to define ourselves by what we are, not by whom we oppose. To use the positive male force is to understand the need for fair and open rules and to take responsibility for upholding them. It is to have the courage to make tough choices and accept responsibility for the results of our actions. It is to act boldly, forthrightly, and do what needs to be done, to leap boldly into the future.

In brief, Berry's resignation letter offers an eloquent articulation of key characteristics of cultural ecofeminism: behavioral traits are inherently gendered; liberation comes from achieving a kind of psychological androgyny, containing a balance of "good" masculine and feminine traits available to both men and women. Berry has explained that her idea of the "Green warrior" is not intended to invoke associations with militarism.[35] Yet this version of cultural ecofeminism is open to exploitation from other forces: in its celebration of traditional gender roles, it does nothing to challenge these dualisms or to make other, nontraditional roles available to women. The female/eros and male/thanatos associations lurk just beneath the surface of Berry's ecofeminism. As socialist feminists first argued, these associations are a product of social construction in the context of an oppressive patriarchal culture; the liberation of women and other oppressed people will come not from reversing the valuation of devalued gender roles but from challenging the very construction of gender itself.

Portions of Berry's letter were repeated verbatim in the "Rationale for Launching a Green Politics Network," published in the March 1992 edition of the *Greens Bulletin*. The GPN founders explain that they had launched a new organization because the original movement "seems unable to come to terms with the oppression of women, and with the positive female and male in all of us." Women have "found the Greens as oppressive as the patriarchal society we are committed to transforming," and men "who have stood up for the positive female have been put down and/or crowded into the old power and control games." In an effort to leave these behaviors behind, GPN founders proposed creating a new organization that valued women's leadership and was capable of providing an alternative to patriarchal politics:

> We must take back our political system from the forces of patriarchy that have defined the nature of politics for the past four thousand years. There is a creative female energy in both men and women at the very depths of our beings that is a strong force for life. This force, like life itself, is sexual, spiritual, messy, ecstatic, joyful, unpredictable, and represents death as well as life. Because patriarchal men fear and distrust this energy in women, in some men (particularly gay men), and in nature itself, they have violently oppressed and dominated this creative energy to the point where all life is now threatened. It is vitally important that all of us, the oppressed and the oppressor alike, liberate ourselves and nature from these patterns of dominance and oppression. This will not be easy. Therefore, because women have almost universally been victims of the oppression and because they embody the life force, they can and must take the lead toward liberation.

Not every woman shared these sentiments, of course. Women in the Greens/Green Party U.S.A. felt quite capable of taking the lead right where they were.

At least three separate statements were written by G/GPUSA women, both individually and collectively, responding in protest to what they perceived as a divisive tactic on the part of the GPN founders. All the women writers believed the GPN statement used the rhetoric of feminism to conceal the GPN's real purpose for leaving the Greens: creating a separate organization for Green Party organizing. Women defending the Greens charged GPN founders with continuing "to organize in a manner that is essentially competitive, antagonistic, and unprincipled." It is significant that none of the women writers challenged the critique of sexism in the Greens. The process of transforming gender relations would be a slow one, women agreed. According to the "Women's Statement on Sexism and Division in the Greens," signed by twelve women, "sexism is pervasive in our society, and it is indeed present in our organization, as in every organization we know of." Stating their commitment to support women and women's leadership in the Greens, the writers asked, "If we abandon hope here, where openness to feminist values is widespread, how can we hope to create change in the larger society?"[36]

With these events in recent memory, women attending the 1992 national Green gathering in Minneapolis made a motion to change the key value from "postpatriarchal values" to "feminism." Arguing that "postpatriarchal" is "difficult to understand" and "implies that our values are coming out of values based on patriarchy," the writers asserted that feminism is both "straightforward and recognizes the rich contributions of feminist theory to the Green movement."[37] The motion did not go forward without considerable opposition; however, a large majority of Greens were ready to acknowledge that sexism was a serious problem within the Greens: "Women are chronically under-represented except where gender balance is mandated," wrote one man. "More profoundly, women have to continually struggle against masculinist styles of work, debate and leadership." In the end, the proposal was adopted by exactly the minimum required (75 percent). Greens hoped its adoption would signal a shift toward "validating and supporting a feminist transformation of The Greens."[38]

Possibly responding to the exodus of several women to the Green Politics Network, members of the Women's Caucus spent a significant portion of their time addressing the importance of women's leadership. One of the ideas to come out of the caucus, along with the key value change to "feminism," was the proposal to implement a Women's Leadership Fund, which would make funds available to women who had reduced access to the funds needed for travel and other expenses related to holding leadership positions. Other proposals for increasing women's leadership included a request to institutionalize women's space at the gatherings and to adjust gender balance on the Greens Coordinating Committee, so that four of the seven seats would be held by women. Unfortunately, none of these proposals addressed the problem of transforming patriarchal behaviors among the Greens, and the problem continued to grow.

That task had to be resolved, mostly, by the men themselves, of course, and many men at the conference challenged other men on patriarchal behaviors.

The new key value received tremendous support from the Men's Caucus, M.E.N. (Men Evolving Naturally), which issued a statement that "feminism is also for men in that it challenges gender stereotypes which impoverish us as well as subordinate women." Resolving to challenge those stereotypes and to act as feminist allies, the Men's Caucus initiated a "Green Sprouts" program to provide childcare at national Green gatherings. The caucus resolution specified that "at least one half of the volunteer shifts assisting with the program shall be filled by men, and at least one half by non-parents."[39] Unlike the Women's Caucus proposal, the Men's Caucus proposal received unanimous support.

Even given these positive actions and four years after "postpatriarchal values" was changed to "feminism," Margaret Garcia, the 1994 California Green Party candidate for secretary of state, lamented, "Whatever happened to Green feminism?" Although she referred to problems specific to the California Green Party, her complaints echoed earlier assessments from women about the Green movement generally. According to Garcia, women were present at the local level, but, higher up, men predominated, and "state party meetings and decisions often reflect the opinions of men who no Green Party locals ever see." In addition, women's views were continually discounted. In a state that began its Green Party "with roughly one man to every three women," Garcia reported, the California Green Party was "lucky if it sees one woman to every seven men." Originally, the California Green Party was known for high numbers of women and for the strength of its women's caucus: "It was there that such physical remedies to male indifference, such as standing on tables, barking, and other behaviors, were born." But women grew tired of making the same points over and over again. "We are accused of bickering when we voice dissent," said one woman. "If you are taking time away from your personal life and children," said another, "you want it to be for a worthwhile project with tangible goals."[40]

In 1996, women in the Green Party of California faced a dilemma they never expected: the presidential campaign of Ralph Nader, initiated by several men in the California Greens. "Our very first chance to run a candidate for president," wrote Garcia, "and we are running a straight, white male who lives three thousand miles from California. So much for growing our own candidates." In Garcia's view, the Nader campaign directly violated "founding promises of introducing only gender-balanced, multi-racial tickets with a woman as head of the ticket." But, like Green women before her, Garcia concluded that such disappointments would only make Green women stronger, more organized and more determined: "We are becoming more assertive among the men in the party because we realize the Green Party of California is not the safe space we thought it would be for feminism." Garcia's frank assessment remained embedded within her essay, where she says: "Do we really think we can have a positive effect on society when we haven't yet adequately dealt with our own sexism and racism?" As Garcia confided a few months after her essay appeared, she was giving the California Green Party until the end of the year; then, if conditions for women had not improved, she, like the other strong women before her, was leaving.[41]

Was the situation for women any different in the Green Politics Network? Surely the declaration that had founded the organization made it seem that women's views would be respected and that women would be encouraged to take leadership roles. In fact, the GPN did feature many women in leadership positions. Ranging in perspectives from "power" feminism (Keiko Bonk) and liberal feminism (Linda Martin, Toni Wurst) to spiritual/cultural ecofeminism (Dee Berry, Anne Goeke), women in the GPN enjoyed support from their male colleagues and a freedom from the debilitating assaults on women's leadership which they felt had characterized the Greens overall.

One cluster of women's leadership in the GPN came from the Hawaii Green Party, which had nurtured the work of Christa Slaton, Keiko Bonk, Toni Wurst, and Linda Martin. In 1992, Keiko Bonk won a seat on the County Council of the Big Island of Hawaii. But Bonk did not attribute her success to feminism: "Feminism is for my mother's generation," she said.[42] To questions about sexism and racism as potential obstacles during her campaign, Bonk has simply responded, "The only instances of sexism were with people instructing me about appropriate attire and makeup. I just told them, 'don't worry about it; my generation can wear lipstick and think at the same time.'" To other women seeking leadership and possibly running for office with the Greens, Bonk counsels, "Be strong and do it. . . . Be bold, be assertive, and do what needs to be done. When people don't feel women can do this or that, it's their problem. Now, we've got to take the power into our own hands."[43] Bonk's strand of feminism is what is popularly called "power feminism," a version of liberal feminism that suggests that if women don't jump into the white male world and take power, it's our own fault.

Nowhere was this position more clear than in Bonk's keynote address to the 1995 Green gathering in Albuquerque. She gave this address after she had been reelected to her position on the Hawaii County Council and gone on to win the chair of that body, after struggling against both Democrats and Republicans. At the Albuquerque gathering, Bonk made it clear that "politics is simply the struggle to use power for specific ends." With statements such as "we are not all equal, and we never will be," and "much of the status-quo is both beautiful and admirable," Bonk thrilled some of her listeners and alienated many others. From a feminist perspective, one of

Bonk's most questionable assertions came in her metaphoric description of politics and the Green official:

> Every Green who gets into office or into a government position supporting someone in office, has to learn one very difficult lesson. They must learn to be polite and civilized while they watch rape. Everyday the powerless and the land are raped by people who do not care, people who enjoy it. And you must watch and smile, because it is your job to get into the middle of the gang rape and look for that one participant that is not real enthusiastic, the one you might be able to influence. You take all the allies you can get, in whatever form they come to you.[44]

From her experience in leadership as an elected Green official, Bonk had learned that "becoming a leader means growing up faster than others," and in her opinion, "it is time for the Greens to grow up."[45] Bonk's use of the maturity metaphor meant a willingness to embrace electoral politics and to work within the system. To those critical of Bonk, the maturity metaphor implied a loss of Green values, particularly the value of the means embodying the ends, and a transition within the Greens and the GPN to sheer political opportunism.

In 1992, another GPN leader, Linda Martin, ran for the U.S. Senate against thirty-year Democratic incumbent Daniel Inouye. This was just months after the Hawaii Green Party qualified for ballot status. Martin received nearly fifty thousand votes (about 14 percent), more than any other third-party congressional candidate in 1992. Although Martin was a first-time candidate, she was no novice. She was prepared for this challenge by her years of work in advertising and marketing, by her background in modeling and her few years as a self-described "corporate wife," by her leadership in a countywide initiative to limit growth in San Diego County, her work to create affordable housing in San Diego and in Hawaii, and by her work with grassroots organizations such as Common Cause/Hawaii. Moreover, her experience writing and producing family-planning films made her uniquely qualified to use the opportunity when serious sexual assault and harassment allegations were lodged against Senator Inouye in the final weeks of the campaign. Promising to keep the issue alive even after the campaign, Martin was instrumental in forming a community-based coalition of more than fifty activists called Code of Silence/Broken. Although Martin's liberal feminism seemed more woman-friendly than Bonk's, both Green women hailed from the school of "power feminism."

Martin took care to present herself as "normal": "I guess when people hear Green Party, they expect someone really radical or something," Martin told Oahu's *Midweek Magazine*. Her carefully crafted self-presentation was part of a larger project of what she saw as "mainstreaming" the Greens: "We are the mainstream of the future," she said. But what was her philosophy? As she told the Oahu reporter, "I wasn't so much attracted to the Greens as I was disgusted with the Democrats."[46] For some Greens, Martin's definition of "mainstream" translated into a kind of reformist politics rather than the more revolutionary politics embraced by grassroots Greens.

Toni Wurst was Martin's campaign co-chair and later took her place as co-chair of the Hawaii Green Party when Martin moved to Virginia. In 1995, Wurst ran for the Hawaii House of Representatives, and in an act of genuine solidarity, Martin returned to Hawaii to spend a month assisting Wurst's campaign. Although Wurst was not elected, she received 41 percent of the vote, an excellent showing for a third-party candidate by any standard. Ideologically, Wurst professed an attraction to the views articulated by ecofeminism but felt it was "too idealistic to be practical"; Green politics, on the other hand, was geared for the real world.[47]

Dee Berry clearly articulated spiritual/cultural ecofeminism, with her approval of "positive" gender traits, her metaphor of the "Green warrior," and her perspectives on Green spirituality. According to Berry, "If we Greens ignore or deny the spiritual basis of our movement, we will not survive."[48]

Reinterpreting and redefining the Ten Key Values, Berry articulated her own Green vision, based on seven values: the politics of empowerment (which relies on grassroots democracy and community-based economics), the politics of satyagraha (or nonviolence, in which she argues for using consensus decision-making processes and opposing militarism), the politics of compassion (meaning social justice and personal and social responsibility), the politics of liberation (or postpatriarchal values), the politics of ecology (or ecological wisdom), the politics of the Green warrior, the politics of joy and celebration ("to have fun and laugh with each other"), and the "acceptance of spirituality as an integral part of all we do, including our politics."[49] Unique to Berry's reformulation are her points about the Green warrior, the politics of joy, and the emphasis on spirituality in politics.

Like Berry, Anne Goeke emphasized embracing "positive" gender traits of the feminine, as explained in her founding pamphlet for the Gylany Greens. "Gylany" is a word taken from Riane Eisler's *The Chalice and the Blade*. Goeke writes that "gyne" means "woman," and "andro" means "man"; the "I" between the two means "linking" rather than "ranking." "In this sense," Goeke explains, "the 'I' stands for the resolution of our problems through the freeing of both halves of humanity from the stultifying and distorting rigidity of roles imposed by the dominating hierarchies inherent in androcratic systems."[50] In the pamphlet, Goeke offers almost two pages of questions. Like Spretnak's original articulation of the Ten Key Values, this list of questions—while admittedly avoiding a rigorous philosophical statement of political beliefs—functions as a way of inviting others to participate in the shaping of answers. Unlike some other ecofeminists, spiritual/cultural ecofeminists seem willing to leave room for different answers.

Goeke belonged to both the G/GPUSA and the GPN because she saw her role as a bridge builder, and spirituality was the link between Green politics and ecofeminism for her. At the 1993 Greens gathering in Syracuse, Goeke and others proposed the Earth Spirituality Caucus as an ongoing project of the G/GPUSA; by 1995, she had also proposed and received endorsement for an Earth Spirituality Network from the GPN. Goeke explained that the purpose of the Earth Spirituality Caucus/Network was to provide an opportunity at the opening and closing of Green gatherings for activists to come together as a community on a spiritual and feeling level to complement and clarify their coming together as political activists. "What we are doing as Greens," Goeke explained, "is more than what happens in our lifetime. If we can just recognize that common bond, take that moment to think about our relationships to each other and to the earth, we can draw on a strength that will carry us through some very difficult moments. If you believe in the interconnectedness of all life—if you have a consciousness of this universal home—then you are believing in a very spiritual aspect of things."[51] From that interconnected sense of self comes a way of valuing self and others based on relationships of care and of survival, which transcends the stereotypically feminine valuing of other more than self and the stereotypically masculine valuing of self more than others. It is also quite different from the deep ecological expansion of the male self to include (erase, annihilate) all others. Goeke, along with other members of the Earth Spirituality Caucus of the G/GPUSA, believed one of the characteristics of earth spirituality is that it promises to "heal the previous antagonism between religion and nature, mind and matter, into one of complementarity and balanced harmony."[52]

Another fairly startling characteristic of spiritual/cultural ecofeminism is its critique of reason. Charlene Spretnak asserts: "To reject the cult of rationalism places thought, feelings, value, ethics, and meaning in the larger context of the Earth community, which is certainly a basis for a coherent politics."[53] Of course, this critique is not the sole property of cultural ecofeminism: as Val Plumwood, an ecofeminist philosopher, has thoroughly explained in *Feminism and the Mastery of Nature*, the "master model" has relied on the definition of reason as the distinguishing characteristic of the master, against which all "others" are defined; lack of "reason" is sufficient justification for subordination. Along similar lines, the Earth Spirituality Caucus members wrote, "Earth Spirituality directs us to collaborate with those in the social activist movements, encouraging them to turn from reason alone as the source of their activities (if this should be the case) to a 'Soul of the Whole.'" As Goeke commented, "Shifting our relationship to the earth, how we perceive the future, how we perceive the past," is the goal of Earth Spirituality, and "that shift is more of an internal shift than an intellectual shift."[54]

For the G/GPUSA, Goeke has served as a representative for the Women's Caucus, the International Working Group, and the Green Global Network. In her community of Lancaster, Pennsylvania, she cofounded a Women in Black group that met weekly on the steps of the county courthouse, holding a silent vigil as a reminder of the Serbian, Muslim, and Croat women who were being raped and beaten in the name of war. She also cofounded the Lancaster Greens in 1990 and organized local events for Earth Day 1995. A board member of the Voice for Choice Coalition, as well as a long-standing member of the National Organization for Women (NOW), the Women's Environment and Development Organization (WEDO), and Co-op America, Goeke has shown her commitment to both Green politics and feminism.

In 1996, the women leaders of the GPN gained additional prominence when Linda Martin established a Draft Nader Clearinghouse to rival the one in California, and Anne Goeke was chosen as Nader's stand-in vice-presidential running mate on the ballots in fourteen states. But the Nader campaign showed the limits of liberal feminism, "power feminism," and spiritual/cultural ecofeminism in addressing the root causes of oppression—and in maintaining a distinct ecofeminist voice or presence in the Greens.

Notes

1. Ynestra King, "Coming of Age with the Greens," *Z Magazine* (February 1988): 19.

2. Charlene Spretnak and Fritjof Capra, *Green Politics* (Santa Fe, N.M.: Bear and Co., 1984), 151.

3. Petra Kelly, "Open Letter to the German Green Party," in *Nonviolence Speaks to Power*, ed. Glenn D. Paige and Sarah Gilliatt (Honolulu: Center for Global Nonviolence Planning Project at the University of Hawaii, 1992), 149–59; Charlene Spretnak, Appended to Petra Kelly's *Thinking Green! Essays on Environmentalism, Feminism, and Nonviolence* (Berkeley, Calif.: Parallax Press, 1994), 151–54.

4. Charlene Spretnak, personal communication, August 23, 1997.

5. Howard Hawkins, "North American Greens Come of Age: Statism vs. Municipalism," *Society and Nature* 3 (1993): 204–205.

6. Charlene Spretnak, ed., *The Politics of Women's Spirituality* (New York: Doubleday, 1982); Spretnak, *The Spiritual Dimension of Green Politics* (Santa Fe, N.M.: Bear and Co., 1986); Mark Satin, *New Age Politics: Healing Self and Society* (New York: Dell, 1978); Satin, *New Options for America: The Second American Experiment Has Begun* (Fresno: California State University Press, 1991).

7. Charlene Spretnak, "Ecofeminism: Our Roots and Flowering," in *Reweaving the World: The Emergence of Ecofeminism*, ed. Irene Diamond and Gloria Feman Orenstein (San Francisco: Sierra Club Books, 1990), 3–14.

8. Spretnak and Capra, *Green Politics*; Satin, *New Age Politics*, 273.

9. Ellen Willis, "Radical Feminism and Feminist Radicalism," in *The Sixties without Apology*, ed. Sohnya Sayres, Anders Stephanson, Stanley Aronowitz, and Fredric Jameson (Minneapolis: University of Minnesota Press, 1984), 91–118.

10. Janet Biehl, "What Is Social Ecofeminism?" *Green Perspectives* 11 (October 1988): 4.

11. Kirkpatrick Sale, "Ecofeminism—A New Perspective," *The Nation* (September 26, 1987): 302–305; Ynestra King, "What Is Ecofeminism?" *The Nation* (December 12, 1987): 702, 730–31.

12. Judith Plant, "Revaluing Home: Feminism and Bioregionalism," in *Home! A Bioregional Reader*, ed. Van Andruss, Christopher Plant, Judith Plant, and Eleanor Wright (Philadelphia: New Society, 1990), 21–23; Judith and Christopher Plant, *Turtle Talk: Voices for a Sustainable Future* (Philadelphia: New Society, 1990).

13. King, "What Is Ecofeminism?" 702.

14. King, "What Is Ecofeminism?" 730–31.

15. Janet Biehl, "An Eco-Feminist Looks at Deep Ecology," *Kick It Over* (special supplement) 20 (winter 1987/88): 2A–4A.

16. Biehl, "Eco-Feminist Looks," 3A–4A.

17. Biehl, "Eco-Feminist Looks," 4A.

18. "Consensus and Community: An Interview with Caroline Estes," in *Healing the Wounds: The Promise of Ecofeminism*, ed. Judith Plant (Philadelphia: New Society, 1989), 235–41.

19. Caroline Estes, telephone interview with author, June 26, 1996.

20. Howie Hawkins, "The Politics of Ecology: Environmentalism, Ecologism, and the Greens," *Resist* 217 (July/August 1989): 6.

21. Estes, telephone interview, June 26, 1996.

22. I am indebted to Charlene Spretnak for copies of these statements and for the attributions that she wrote for each section or statement.

23. Paul Stark, "Why Detroit Summer?" *Green Letter* 7 (fall 1991): 2–3.

24. Chaia Heller, "Down to the Body: Feminism, Ecology, and the Evolution of the Body Politic," *Society and Nature* 2 (1993): 142.

25. First published in *Renewing the Earth*, ed. John Clark (London: Green Print, 1990), Heller's essay "Toward a Radical Ecofeminism: From Dua-Logic to Eco-Logic" is available to U.S. readers in the journal *Society and Nature* 2, 1 (1993): 72–96. Heller has also published an essay, "For the Love of Nature: Ecology and the Cult of the Romantic," in *Ecofeminism: Women, Animals, Nature*, ed. Greta Gaard (Philadelphia: Temple University Press, 1993), 219–42. Most recently, she has developed her theory of social ecofeminism in *The Revolution That Dances* (Aigis Publications, forthcoming).

26. Kate Fox and Paul Glavin, "The Politics of Imagination and Struggle," *Record of Antioch College* 44 (June 2, 1989): 8–9.

27. April Cope, "Heller, Eco-Feminists Conflict with Greens," *Record of Antioch College* 46 (June 2, 1989): 5.

28. For more on the Women's Pentagon Actions see Ynestra King, "If I Can't Dance in Your Revolution, I'm Not Coming," in *Rocking the Ship of State: Toward a Feminist*

Peace Politics, ed. Adrienne Harris and Ynestra King (Boulder, Colo.: Westview Press, 1989), 281–98.

29. "Women's Caucus Report," *Left Green Notes* 3 (July/August 1990): 17.

30. Maura Dillon, "Anarchism, Feminism, and Building a Movement," *Left Green Notes* 5 (November/December 1990): 8–9, 11.

31. Janet Biehl, *Rethinking Ecofeminist Politics* (Boston: South End Press, 1991).

32. The favorable review of Biehl's book, written by Kelly Stoner, appears in *Left Green Notes* 7 (April/May 1991): 11. Transcripts of Schere's and Sandilands' presentations, along with the revised "women's liberation" principle, were published in *Left Green Notes* 9 (August/September 1991); see Kate Sandilands, "They Don't Call Us a Movement for Nothing: A Preface to the (New) Left Green Principle on Women's Liberation," 29; Laura Schere, "Feminism, Ecology, and Left Green Politics," 30–31.

33. Howie Hawkins, "The Green Gathering in Elkins, West Virginia: You Should've Been There," *Left Green Notes* 10 (November/December 1991): 6–18.

34. Christa Daryl Slaton, "The Failure of the United States Greens to Root in Fertile Soil," 83–117, in *Research in Social Movements, Conflicts, and Change: The Green Movement Worldwide,* ed. Matthias Finger (London: JAI Press, 1992), 83–117.

35. Dee Berry, telephone interview with author, June 11, 1996.

36. "Women's Statement on Sexism and Division in the Greens," *Greens Bulletin* (March 1992): 47.

37. Ann Poland, Aspen Olmstead, Martha Percy-Meade, Syracuse Greens, "Change 'post-Patriarchal Values' to 'Feminist Values,'" *Green Tidings* 4 (August 8, 1992): 9.

38. Joseph Boland, Amy Belanger, Diana Spalding, and David Shlosberg, "Healing Rifts: Community and Organization in the Fifth Greens Gathering," *Groundwork* (fall 1992): 48–51.

39. Diana Spalding, "The Greens National Gathering," *Z Magazine* 5 (November 1992): 55.

40. Margaret Elysia Garcia, "Whatever Happened to Green Feminism?" *Z Magazine* 9 (March 1996): 20–21.

41. Garcia, "Whatever Happened?" 20–21; Margaret Garcia, telephone interview with author, May 1996.

42. Keiko Bonk, personal interview with author, May 23, 1993.

43. "Keiko Bonk-Abramson: A New Green Officeholder Interviewed on Her Campaign and Green Justice," *Green Politics* 2 (spring 1993): 10.

44. A transcript of Keiko Bonk's keynote was printed as "Growing Up Politically: The Courage to Act," in the GPN newsletter, *Green Horizon* 4 (October/November 1995): 1–5.

45. Bonk, "Growing Up Politically," 5.

46. Don Chapman, "A Still-Green Politician: Hawaii Greens Co-Chair Linda Martin Is New to Politics, but She Brought Recognition to Her Party," *Midweek Magazine* 9 (November 11, 1992): A10.

47. Toni Wurst, personal interview with author, May 27, 1993.

48. Dee Berry, *The Challenge of the Greens: Making the Spiritual-Political Connection* (pamphlet, February 12, 1989), 25.

49. Dee Berry, *A Green Story* (pamphlet, circa 1988), 13–14.

50. Riane Eisler, *The Chalice and the Blade* (San Francisco: HarperCollins, 1987); Anne Goeke, "A Way of Being GyIany Green" (pamphlet, n.d.).

51. Ann Goeke, telephone interview with author, July 8, 1996.

52. Earth Spirituality Caucus, "Greens USA Integrate Earth Spirituality," *Greens Bulletin* (June/July 1994): 16–17.

53. Charlene Spretnak, "Improving the Debates," *Green Synthesis* 33 (March 1990): 5.

54. Earth Spirituality Caucus, "Greens USA Integrate," 17; Goeke, telephone interview, July 8, 1996.

Selected Bibliography

Adams, Carol J., ed. *Ecofeminism and the Sacred*. New York: Continuum, 1994.

Agenda 21: An Easy Reference to the Specific Recommendations on Women. New York: UNIFEM, undated.

Barlow, Maude, and Tony Clarke. *Global Showdown: How the New Activists Are Fighting Global Corporate Rule*. Toronto: Stoddart, 2001.

Berry, Thomas. *The Dream of the Earth*. San Francisco: Sierra Club Books, 1988.

Biehl, Janet. *Finding Our Way: Rethinking Ecofeminist Politics*. Montreal: Black Rose Books, 1991.

Bila, Sorj. "O feminismo como metafora da natureza." *Revista de Estudos Feministas*. CIEC Escola de Comunicao UFRJ, n.092.

Birch, Charles, William Eakin, and Jay B. McDaniel, eds. *Liberating Life: Contemporary Approaches to Ecological Theology*. Maryknoll, N.Y.: Orbis Books, 1990.

Blunt, Alison, and Gilliam Rose, eds. *Writing Women and Space: Colonial and Postcolonial Geographies*. New York: Guilford, 1994.

Braidotti, Rosi, Ewa Charkiewicz, Sabine Haüsler, and Saski Wieringa. *Women, the Environment and Sustainable Development: Towards a Theoretical Synthesis*. London and Atlantic Highlands, N.J.: Zed, 1994.

Brock, Rita Nakashima. *Journeys by Heart: A Christology of Erotic Power*. New York: Crossroad, 1991.

Brown, Joanne Carlson, and C. Bohn, eds. *Christianity, Patriarchy and Abuse*. Cleveland: The Pilgrim Press, 1989.

Callicott, J. Baird. *Earth's Insights: A Survey of Ecological Ethics from the Mediterranean Basin to the Australian Outback*. Berkeley: University of California Press, 1994.

Caswell, Tricia. "Australia and Asia—The Environmental Challenge." In *Living with Dragons: Australia Confronts its Asian Destiny*, ed. Greg Sheridan. St. Leonards, New South Wales: Allen and Unwin, 1995.

Chopp, Rebecca, and Sheila Greeve Davaney, eds. *Horizons in Feminist Theology: Identity, Tradition and Norms.* Minneapolis: Fortress, 1997.

Chuan-Dao, ed. *Buddhism & Social Concern—Life, Ecology & Environmental Concern.* Taipei: Modern Buddhists Association, 1994.

Clark, John, ed. *Renewing the Earth.* London: Green Print, 1990.

Clarke, Robert. *Ellen Swallow: The Woman who Founded Ecology.* Chicago: Follet, 1973.

Clarke, Tony. *Silent Coup: Confronting the Big Business Takeover of Canada.* Toronto: James Lorimer and Co., 1997.

Colborn, Theo, Dianne Dumanoski, and John Peterson Myers. *Our Stolen Future.* New York: Plume, 1997.

Con-spirando: Revista latinoamericana de ecofeminismo, espiritualidad y teología. Santiago de Chile, Chile: Con-spirando Collective, 1992.

Crosby, Donald, and Charley Hardwick, eds. *Religious Experience and Ecological Responsibility.* New York: Peter Lang, 1996.

Cuomo, C. J. "Unraveling the Problems in Ecofeminism." *Environmental Ethics* 14, no. 4 (1992): 351–63.

d'Eaubonne, Françoise. *Le Feminisme ou la Mort.* Paris: Pierre Horny, 1974.

Darcy de Oliveira, Rosiska, and Thais Corral. *Terra Femina.* Rio de Janeiro: A Joint Publication of Institute for Cultural Action (IDAC) and The Network in Defense of Human Species (REDEH), 1992.

Diamond, Irene, and Gloria Feman Orenstein, eds. *Reweaving the World: The Emergence of Ecofeminism.* San Francisco: Sierra Club Books, 1990.

Dobbin, Murray. *The Myth of the Good Corporate Citizen: Democracy Under the Rule of Big Business.* Toronto: Stoddart, 1998.

Dobson, Andrew. *Green Political Thought,* 2d ed. London: Routledge, 1995.

Dobson, Andrew, and Paul Lucardie, eds. *The Politics of Nature.* London: Routledge, 1993.

Dussel, Enrique. *Teología de la Liberación y Etica.* Buenos Aires: Latinoamérica Libros, 1974.

——. *Etica Comunitaria. Teología Liberta Collection.* Petropolis: Ed. Vozes, 1987.

Eaton, Heather. "At the Intersection of Ecofeminism and Religion: Directions for Consideration." *Ecotheology* 11 & 23 (2002): 91–107.

——. "Ecofeminism and Theology: Challenges, Confrontations and Reconstructions." In *Christianity and Ecology: Wholeness, Respect, Justice, Sustainability.* Ed. Dieter Hessel and Rosemary Radford Ruether. Cambridge: Harvard Center for the Study of World Religions, 2000.

Eber, Christine. *Women and Alcohol in a Highland Maya Town: Water of Hope, Water of Sorrow.* Austin: University of Texas Press, 1995.

Eisler, Riane. *The Chalice and the Blade.* San Francisco: HarperCollins, 1987.

Freire, Paolo. *Pedagogy of the Oppressed.* New York: Herder and Herder, 1970.

Fulkerson, Mary McClintock. *Changing the Subject: Women's Discourses and Feminist Theology.* Minneapolis: Fortress, 1994.

Gaard, Greta, ed. *Ecofeminism: Women, Animals, Nature.* Philadelphia: Temple University Press, 1993.

Gebara, Ivone. *Teologia Ecofeminista.* Sao Paulo: Olho dígua, 1998.

———. *Longing for Running Water: Ecofeminism and Liberation.* Minneapolis: Fortress Press, 1999.

Goldenberg, Naomi. *The Changing of the Gods: Feminism and the End of Traditional Religions.* Boston: Beacon Press, 1979.

Griffin, Susan. *Woman and Nature: The Roaring Inside Her.* New York: Harper and Row, 1978.

Haraway, Donna. *Modest_Witness@Second_Millennium.FemaleMan©_Meets_ OncoMouse™.* New York: Routledge, 1997.

Harrison, Beverly Wildung, and Carol Robb, eds. *Making the Connections: Essays in Feminist Social Ethics.* Boston: Beacon, 1985.

Harrison, Bruce. *Going Green: How to Communicate Your Company's Environmental Commitment.* Chicago: Irwin Professional Publishers, 1993.

Heng-Ching. *Wonderful Women in the Bodhisattva's Way.* Taipei: Dong Da, 1995.

Herbert, J. *Shinto: At the Fountain-Head of Japan.* London: Allen and Unwin, 1967.

Hernández Castillo, Rosalva Aida, ed. *La Otra Palabra: mujeres y violencia en Chiapas, antes y después de Acteal.* San Cristóbal de Las Casas, Chiapas, México: Centro de Investigaciones y Estudios Superiores, en Colectivo de Encuentro entre Mujeres (COLEM) y el Centro de Investigación y Acción para la Mujer (CIAM), 1997.

Johnson, Elizabeth. *She Who Is: The Mystery of God in Feminist Theological Discourse.* New York: Crossroad, 1993.

Karliner, Joshua. *The Corporate Planet: Ecology and Politics in the Age of Globalization.* San Francisco: Sierra Club, 1997.

Kaza, Stephanie. "Acting With Compassion: Buddhism, Feminism, and the Environmental Crisis." In *Ecofeminism and the Sacred,* ed. Carol J. Adams. New York: Continuum, 1994.

Keck, Margaret E., and Kathryn Sikkink. *Activists beyond Borders.* Ithaca and London: Cornell University Press, 1998.

Kelly, Petra. *Fighting for Hope.* London: Chatto and Windus, 1984.

———. *Thinking Green! Essays on Environmentalism, Feminism, and Nonviolence.* Berkeley: Parallax Press, 1994.

King, Ynestra, and Adrienne Harris, eds. *Rocking the Ship of State: Toward a Feminist Peace Politics.* Boulder, Colo.: Westview Press, 1989.

Kittay, Eva Feder, and Diana Meyers, eds. *Women and Moral Theory.* Lanham, Md.: Rowman & Littlefield, 1987.

Klein, Naomi. *No Logo: Taking Aim at the Brand Bullies.* Toronto: Vintage, 2000.

Korten, David. *When Corporations Rule the World.* West Hartford, Conn.: Kumarian Press, 1995.

———. *The Post Corporate World: Life after Capitalism.* West Hartford, Conn.: Kumarian Press, 1999.

Leach, Melissa. *Rainforest Relations: Gender and Resource Use Among the Mende of Gola, Sierra Leone.* Washington, D.C.: Smithsonian Institution Press, 1994.

Lewenhak, Sheila. *The Revaluation of Women's Work.* London: Earthscan, 1992.

Li, F., and R. Ju., eds. *Gender, Spirituality & Taiwanese Religions.* Taipei: Academia Sinica, 1997.

Lorentzen, Lois Ann. "Reminiscing about a Sleepy Lake: Borderland Views of Place, Nature/Culture from a Salvadoran Context." In *Wild Ideas,* ed. David Rothenberg. Minneapolis: University of Minnesota Press, 1995.

——. "Women, the Environment and Sustainable Development: Cases from Central America." In *Ecological Resistance Movements: The Global Emergence of Radical and Popular Environmentalism,* ed. Bron Taylor. Albany, N.Y.: SUNY Press, 1995.

Lorentzen, Lois Ann, and Jennifer Turpin, eds. *The Women and War Reader.* New York: New York University Press, 1998.

MacKinnon, Mary Heather, and Moni McIntyre, eds. *Readings in Ecology and Feminist Theology.* Kansas City, Mo.: Sheed and Ward, 1995.

McQuaig, Linda. *All You Can Eat: Greed, Lust and the New Capitalism.* Toronto: Penguin, 2001.

Mellor, Mary. *Breaking the Boundaries: Towards a Feminist Green Socialism.* London: Virago, 1992.

——. *Feminism & Ecology.* New York: New York University Press, 1997.

Merchant, Carolyn. *The Death of Nature.* New York: Harper and Row, 1983.

——. *Earthcare: Women and the Environment.* New York: Routledge, 1995.

Mies, Maria. *Patriarchy and Accumulation on a World Scale.* London: Zed, 1986.

Mies, Maria, Veronika Bennholdt-Thomson, and Claudia von Werlhof. *Women: The Last Colony.* London and Atlantic Highlands, New Jersey: Zed, 1988.

Mies, Maria, and Vandana Shiva. *Ecofeminism.* London: Zed, 1993.

Morin, Edgar. *Le paradigme perdu: la nature humaine.* Paris: Éditions du Seuil, 1973.

Myers, N. *The Primary Source: Tropical Forests and Our Future.* New York: W. W. Norton, 1984.

Nivedita, Menon, ed. *Gender and Politics in India.* Delhi: Oxford University Press, 1999.

Nyamweru, Celia. "Sacred Groves Threatened by Development." *Cultural Survival Quarterly* (fall 1996): 19–21.

O'Connor, Martin, ed. *Is Capitalism Sustainable? Political Economy and the Politics of Ecology.* New York: Guilford Press, 1994.

Parkin, David. *Sacred Void: Spatial Images of Work and Ritual among the Giriama of Kenya.* Cambridge: Cambridge University Press, 1991.

Plant, Judith, ed. *Healing the Wounds: The Promise of Ecofeminism.* Philadelphia: New Society, 1989.

——. "Revaluing Home: Feminism and Bioregionalism." Pp. 21–23 in *Home! A Bioregional Reader,* ed. Van Andruss, Christopher Plant, Judith Plant, and Eleanor Wright. Philadelphia: New Society, 1990.

Plant, Judith, and Christopher Plant. *Turtle Talk: Voices for a Sustainable Future.* Philadelphia: New Society, 1990.

Plumwood,Val. *Feminism and the Mastery of Nature.* London: Routledge, 1993.

Ramachandra, Guha. *Social Ecology.* Delhi: Oxford University Press, 1998.

Rasmussen, Larry. *Earth Community, Earth Ethics.* Maryknoll, N.Y.: Orbis Books, 1996.

Reinharz, Shulamit. *Feminist Methods in Social Research.* New York: Oxford University Press, 1992.

Rocheleau, Dianne, Barbara Thomas-Slayter, and Esther Wangari, eds. *Feminist Political Ecology: Global Issues and Local Experiences.* London and New York: Routledge, 1996.

Rosaldo, Michele Zimbalist, and Louise Lamphere. *Women, Culture, and Society*. Palo Alto, Calif.: Stanford University Press, 1974.

Rosenbaum, Brenda. *With Our Heads Bowed: The Dynamics of Gender in a Maya Community*. Albany, N.Y.: SUNY Press, 1993.

Rowell, Andrew. *Green Backlash: Global Subversion of the Environmental Movement*. London: Routledge, 1996.

Roy, Arundathi. *The Greater Common Good*. Bombay: India Book Distributors Ltd., 1999.

Ruether, Rosemary Radford. *New Woman, New Earth*. New York: The Seabury Press, 1975.

———. *Gaia & God, An Ecofeminist Theology of Earth Healing*. San Francisco: Harper, 1992.

———. *Women Healing Earth: Third World Women on Ecology, Feminism and Religion*. Maryknoll, N.Y.: Orbis Books, 1996.

Salleh, Ariel. *Ecofeminism as Politics: Nature, Marx and the Postmodern*. London and New York: Zed Books Ltd., 1997.

Satin, Mark. *New Options for America: The Second American Experiment Has Begun*. Fresno: California State University Press, 1991.

Seager, Joni. *Earth Follies: Coming to Feminist Terms with the Global Environmental Crisis*. New York: Routledge, 1993.

Sen, Gita, and Caren Grown. *Development Crises and Alternative Visions*. New York: Monthly Review Press, 1987.

Shiva, Vandana. *Staying Alive: Women, Ecology and Development*. London: Zed, 1989.

———. *Biopiracy: The Plunder of Nature and Knowledge*. Toronto: Between the Lines, 1997.

———. *Reith Lecture* 2000. http://news.bbc.co/uk/hi/english/static/vents/reith_2000lecture5.stm

———, ed. *Minding our Lives: Women from the South and North Reconnect Ecology and Health*. Delhi: Kali for Women, 1993.

———, ed. *Close To Home: Women Reconnect Ecology, Health and Development Worldwide*. Philadelphia: New Society Publishers, 1994.

Silverblatt, Irene. *Moon, Sun and Witches: Gender Ideologies and Class in Inca and Colonial Peru*. Princeton, N.J.: Princeton University Press, 1987.

Slaton, Christa Daryl. "The Failure of the United States Greens to Root in Fertile Soil." Pp. 83–117 in *Research in Social Movements, Conflicts, and Change: The Green Movement Worldwide*, ed. Matthias Finger. London: JAI Press, 1992.

Sölle, Dorothee. *The Window of Vulnerability: A Political Spirituality*. Minneapolis: Fortress, 1990.

Spretnak, Charlene. *The Spiritual Dimension of Green Politics*. Santa Fe, N.M.: Bear and Co., 1986.

———. *States of Grace*. New York: Harper Collins, 1991.

———, ed. *The Politics of Women's Spirituality*. New York: Doubleday, 1982.

Spretnak, Charlene, and Fritjof Capra. *Green Politics*. Santa Fe, N.M.: Bear and Co., 1984.

Sturgeon, Noël. *Ecofeminist Natures: Race, Gender, Feminist Theory and Political Action*. London and New York: Routledge, 1997.

Támez, Elsa. "Latin American Feminist Hermeneutics: A Retrospective." In *Women's Visions: Theological Reflection, Celebration, Action*, ed. Ofelia Ortega. Geneva: WCC Publications, 1995.

Tobias, Michael, ed. *Deep Ecology*. San Marcos, Calif.: Avant Books, 1984.

Tsing, Anna. "Environmentalisms: Transitions as Translations." In *Transitions, Translations, Environments: International Feminism in Contemporary Politics*, ed. Joan Scott, Cora Kaplan, and Debra Keates. New York: Routledge, 1997.

Turpin, Jennifer, and Lois Ann Lorentzen, eds. *The Gendered New World Order: Militarism, the Environment, Development*. New York: Routledge, 1996.

van den Hombergh, Heleen. *Gender, Environment and Development: A Guide to the Literature*. Utrecht: International Books, 1993.

Waring, Marilyn. *If Women Counted: A New Feminist Economics*. San Francisco: Harper Collins, 1988.

Warren, Karen J. *Ecological Feminist Philosophies*. Bloomington: University of Indiana Press, 1996.

———. *Ecofeminist Philosophy: A Western Perspective on What it Is and Why it Matters*. Lanham, Md.: Rowman & Littlefield, 2000.

———, ed. *Ecological Feminism*. London and New York: Routledge, 1994.

———, ed. *Ecofeminism: Women, Culture, Nature*. Bloomington: Indiana University Press, 1997.

Williams, Delores. *Sisters in the Wilderness: The Challenge of Womanist God-Talk*. New York: Orbis Books, 1993.

Willis, Ellen. "Radical Feminism and Feminist Radicalism." Pp. 91–118 in *The Sixties without Apology*, ed. Sohnya Sayres, Anders Stephanson, Stanley Aronowitz, and Fredric Jameson. Minneapolis: University of Minnesota Press, 1984.

Index

About the Contributors

Heather Eaton is a professor at Saint Paul University, Canada, in the areas of feminism, ecology, and religion. She is a socially engaged academic and frequently participates in public education and social activism. Her interests lie in interreligious responses to the ecological crisis, particularly the relationship between ecological, feminist, and liberation theologies. She has published numerous chapters in books and articles in such journals as *Ecotheology*, *Worldviews*, *Journal of Feminist Theology*, and *Theoforum*. She is a member of the American Academy of Religion, the Canadian Theological Society, Environmental Studies Association of Canada, and Women, Environment, Education, and Development.

Greta Gaard has been an associate professor of women's studies and composition at the University of Minnesota-Duluth and an associate professor of humanities at Fairhaven College, Western Washington University. She is completing a book of ecofeminist, creative nonfiction essays titled *Home Is Where You Are*. Greta's edited volumes include *Ecofeminism: Women, Animals, Nature* and *Ecofeminist Literary Criticism: Theory, Interpretation and Pedagogy*.

Ivone Gebara is a Brazilian writer who works on themes of philosophy and theology from a feminist and ecological perspective. Her books include *Teologia Ecofeminista* and *Longing for Running Water: Ecofeminism and Liberation*. For the last seventeen years she has been a professor at the Theological Institute of Recife, Brazil. At present, she is a visiting professor at a variety of universities, both in Brazil and abroad. She also gives talks and workshops to women in grassroots communities in the northeast of Brazil.

Aruna Gnanadason coordinates the Women's Programme, as well as the Justice, Peace, and Creation Team of the World Council of Churches (WCC) in Geneva. She authored *No Longer a Secret: The Church and Violence Against Women*, published by the WCC, and has contributed many articles to Christian and secular journals, magazines, and books. She is a member of the Ecumenical Association of Third World Theologies. Gnanadason received honorary doctorates from both the Academy of Indian Ecumenical Theology and Church Administration and the Senate of Serampore Colleges, India. Presently she is a candidate for a doctoral degree in ministries in the area of feminist theologies with the San Francisco Theological Seminary in the United States.

Wan-Li Ho teaches Chinese and religion at Emory University. Dr. Ho received her Ph.D. in religion from Temple University, where she specialized in women in religion and environmental issues. Chinese religions, comparative thought, Asian women's studies, and ecofeminism have long been serious academic interests of hers. She has written or co-authored a number of articles and books on these themes, including *The Tao of Jesus: An Experiment in Inter-Religious Understanding* (New York: Paulist Press, 1998), co-authored with Father Joe Loya and Chang-Shin Jih.

Lois Ann Lorentzen is professor of social ethics, associate director of the Center for Latino Studies in the Americas, and principal investigator for the Religion and Immigration Project at the University of San Francisco. She is co-chair of the Religion in Latin America and the Caribbean Group of the American Academy of Religion and past chair of the Ecology section of the American Academy of Religion—Western Region. She is editor or author of *Ética Ambiental* (Environmental Ethics); *The Gendered New World Order: Militarism, Environment, Development; The Women and War Reader; Liberation Theologies, Postmodernity and the Americas; Religions/Globalization: Theories and Cases.*

Masatsugu Maruyama is a professor in political theory at Yamanashigakuin University, Japan. He is co-author of *Political Studies of "New Politics"* and *Invitation to Green Politics*, both published in Japan. His work combines democracy theory with Green politics issues. He is currently completing a new book entitled *Green Political Theory as the Reflection of Modernity.*

Mary Mellor is a professor in the School of Arts and Social Sciences at Northumbria University, UK, where she is also chair of the Sustainable Cities Research Institute. Her current research is on sustainable economics drawing on ecofeminist and social economy principles. Her most recent book, jointly authored with Frances Hutchinson of Bradford University and Wendy Olsen of Manchester University, is *The Politics of Money: Towards Sustainability and Economic Democracy* (Pluto Press, 2002).

Celia Nyamweru is an associate professor of anthropology at St. Lawrence University in Canton, New York. She was born in London, England, where she earned her degrees in geography from Cambridge University. She first traveled to Kenya in 1965. Subsequently she taught at Kenyan universities for twenty years. She has done research in Kenya and other parts of Africa on a wide array of topics ranging from paleoclimatology to traditional agriculture to conservation of natural sacred sites.

Mary Judith Ress, a native of the United States, has lived in Latin America since 1970. She has masters degrees in economics, theology, and Spanish language and literature and is currently completing work on her doctorate in feminist theology at the San Francisco Theological Seminary. She was managing editor of Latinamerica Press/Noticias Aliadas (1981 to 1989) and editor of the bi-monthly IDOC Internacionale in Rome (1989–1990). She is currently a founding editor of *Con-spirando: Revista latinoamericana de ecofeminismo, espiritualidad y teología,* a quarterly journal published in Santiago, Chile. She is the author of *Lluvia para florecer: Entrevistas sobre el ecofeminismo en América Latina* (Santiago, Chile: Colectivo Con-spirando, 2002). She is married to David Molineaux and mother to Peter (24) and Benjamin (23).

Noël Sturgeon is associate professor and chair of women's studies as well as graduate faculty in American studies at Washington State University. She has a Ph.D. from the History of Consciousness Program at the University of California, Santa Cruz (UCSC). She has been a visiting ecofeminist scholar at Murdoch University in Perth, Australia, visiting professor at the JFK Institute for North American Studies at the Free University in Berlin, Germany, a fellow of the Center for Cultural Studies, UCSC, and a Rockefeller fellow at the Center for the Critical Analysis of Contemporary Culture at Rutgers University. She is the author of *Ecofeminist Natures* (Routledge, 1997) and numerous articles on ecofeminism, direct action movements, and environmental cultural studies.